IBRA

**INTERNATIONAL BEE
RESEARCH ASSOCIATION**

www.ibrabee.org.uk

University of
Salford
MANCHESTER
1967-2017 50 YEARS

www.northernbeebooks.co.uk

The Asian Hornet *(Vespa velutina)* Threats, Biology & Expansion

by

Professor Stephen John Martin
(University of Salford, Manchester, UK)

ISBN: 978-0-86098-281-4

Jointly published by:

The International Bee Research Association, A Company limited by Guarantee, 91 Brinsea Road, Congresbury, Bristol, BS49 5JJ (UK) & Northern Bee Books, Scout Bottom Farm Mytholmroyd, Hebden Bridge HX7 5JS (UK)

Obtainable from:

www.ibrabee.org.uk & www.northernbeebooks.co.uk

The Asian Hornet
(*Vespa velutina*)

Threats, Biology & Expansion

by

Professor Stephen John Martin
(University of Salford, Manchester, UK)

*A Yellow hornet (*Vespa simillima*) with a male honey bee in its mandibles that it has just caught. Photograph by S. Martin.*

And I will send hornets before you, which shall drive out the Hivites, the Canaanites, and the Hittites from before you.

Exodus 23:28

*An oriental hornet (*Vespa orientalis*) in Greece ready to attack. Photograph J. Phipps.*

Acknowledgements

I must thank all the people that have helped me during the location, collection and study of around 500 hornet nests. They are too numerous to mention but special mention must go to the Japanese hornet hunter Tatashi-san, my mentors Dr. M. E. Archer and Dr. S. Yamane and finally my wife Yuko, for a life-time of support that made these studies possible (P.S. my kids were of no help at all!). Special thanks go to Mervyn Eddie, Susan Martin and Grant Elliot for proofing the text, so any mistakes blame them and finally all the people that freely contributed their photographs, especially Chris Luck, John Phipps, Yoshiaki Sakai and Junichi Takahashi. Without their and others (see photo credits) generous support this book would be far less interesting. I have attempted to ensure that the information contained in this book is as accurate as possible at the time of going to press. However, in order to make the information more accessible to non-specialists some information has been generalised and those wanting to really get into the details are advised to read the more scientific books or peer review papers provided at the end of the book.

Contents

Preface: My first encounter with hornets 12

Chapter 1. GENERAL INTRODUCTION TO HORNETS & YELLOW-JACKETS

Hornets & wasps, so what is the difference? 17

Hornets and yellow-jackets of the world...21

Hornets and yellow-jackets of Europe 23

Chapter 2. HORNET & YELLOW-JACKET INVASIONS

Accidental introductions of wasps and hornets around the world 25

What traits helps wasps to invade...26

The Asian hornet and its European cousins...29

The Asian hornets' arrival in France and subsequent spread 31

The Asian hornets' arrival in the UK...33

Was the Asian hornet eradicated from the UK? 35

The Asian hornets' arrival in South Korea and subsequent spread 35

Chapter 3. HORNET LIFE CYCLE

Introduction 39

Hibernation (Oct/Nov to April)..39

Post-hibernation migrations...42

The embryo nest (April/May-June), solitary period 44

Protection of the embryo nest from ants 47

Queen usurpation 48

Cooperative period (June)...................................49

The worker nest (June-August) or polyethic period 50

Colony relocation...................................54

Reproductive phase (September-October) 57

Mating...................................61

Colony decline phase (October-November)...................................61

Nest Thermoregulation 62

Diet 66

Chapter 4. HORNET PREDATORS & PESTS

Predators of Hornets 69

Internal parasites of Hornets...................................71

Chapter 5. HORNETS AS PESTS, FOOD & CONTROL METHODS

Introduction 77

Hornets attacking Eastern honey bees...................................78

Hornets attacking Western honey bees 80

Hornets as stinging insects...................................85

Hornets as pests of fruit crops 88

Hornets as food for humans 89

Hornet control methods...................................90

Protecting your honey bee colonies...................................93

The future 95

MORE INFORMATION

Further reading (books)...................................99

Further reading (scientific papers)...................................99

Photo credits

Front cover: Asian hornets hawking, Junichi Takahashi

Back cover: Asian hornet head, Yoshiaki Sakai

Front piece: nest on Buddha, SJM

End of contents page: an oriental hornet, John Phipps

End piece: two photos of Asian hornet hawking, Yoshiaki Sakai

Photographs contributed freely by the generosity of the following people:

George Barker

Tristram Breistaff Creative Commons Attribution-Share Alike 3.0 unported

Africa Gómez (www.abugblog.blogspot.co.uk)

Chris Luck (www.planetepassion.eu)

John Phipps

Stephen Martin (author)

Masato Ono, Tamagawa University, Tokyo, Japan

Lynton Naiff

Yoshiaki Sakai

Katsuhiko Sayama, Forest Products Research Institute, Hokkaido, Japan.

Junichi Takahashi, Kyoto Sangyo University, Japan

Yōuichi Tatashi

Tatsumi Yamamura

Preface:
My first encounter
with hornets

This was on a cold autumn day in the remote Japanese Alps back in 1987, there I discovered a massive hornet nest, just under a meter in diameter, attached to the underside of a concrete road bridge. As the air temperature was around 5°C and all insects are cold blooded, I approached the nest closely to admire its beautiful structure. However, when I tried to obtain a piece of the exquisitely designed nest envelope I was suddenly and unexpectedly attacked by several large yellow hornets that successfully chased me away. Perplexed, I returned the next evening armed with a thermometer (typical scientist), which I plunged into the nest and ran away, again being chased by some angry hornets. Then from a safe distance using a pair of binoculars I observed the temperature inside the nest rise to an amazing 30°C! Had I discovered the first warm-blooded insects? Would I get a Nobel Prize for my discovery? Well obviously-not, thermoregulation in wasps had already been described way back in 1932 by the German scientist Bernard Himmer, and as every beekeeper knows, their bees are able to carefully control the temperature within their colony. Still excited and undeterred I returned and placed more thermometers into the nest (Figure 1) and started to track the nest temperature reported in my first ever scientific publication on hornets entitled 'Thermoregulation in *Vespa simillima xanthopthera*', a riveting read. So, although no Nobel Prize awaited, it did introduce me to the wonderful world of social insects, a subject that I continue to find absolutely fascinating and has over the years become my life's work.

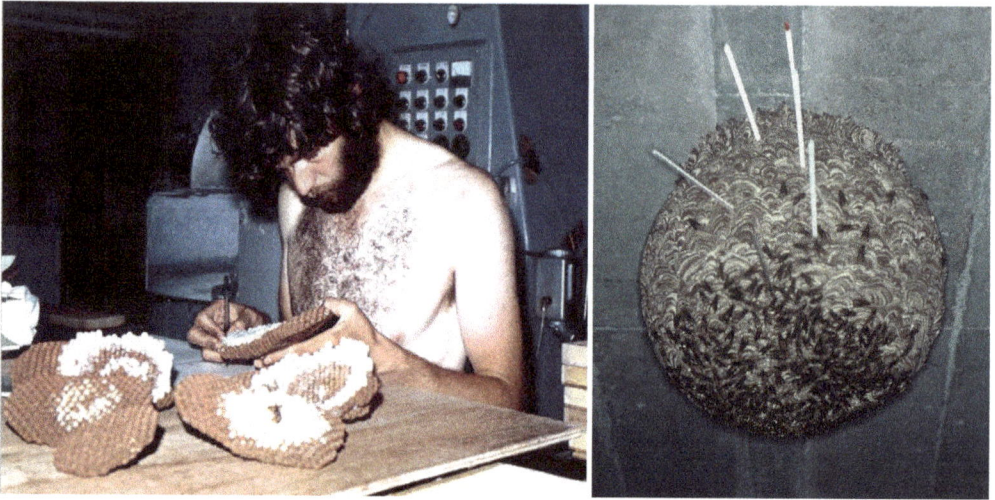

Figure 1. Me as a young Masters student studying my first hornet nest (left) that I discovered under a road bridge into which I placed several thermometers (right) at great personal risk. Photographs by S Martin.

Although my original postgraduate research scholarship to Japan was to study high altitude beetles, this painful encounter with a hornet colony, led to most of the next seven years I spent in Japan being dedicated to understanding the population dynamics and natural history of hornets, from up in the mountains of central Japan, down to the remote sub-tropical Japanese islands (Ogasawara islands) near Taiwan. Much of the basic information provided in this book is based on my studies into various Japanese hornet species, especially the yellow hornet (*Vespa simillima*), a species that is a very similar size and has the same nesting biology as the Asian hornet. In addition, I spent years conducting post-doctoral studies on several other species of hornets, including the Asian hornet all over South West Asia. Over the years, I have removed for study well over 500 hornet colonies, including the Japanese giant hornet - the world's largest species. I have supplemented this information with the increasing amount of scientific literature on the Asian hornet, personal experiences and knowledge gained from working alongside many hornet experts, especially the hornet hunters of the Japan.

Figure 2. Yuichi Tatashi a Japanese 'hornet hunter' from the central mountainous region, holding a mature yellow hornet colony in late autumn. The odd alliance between myself and these people allowed me to collect sufficient data for my research and them to increase the number of colonies they collected and ate. Photograph by S Martin.

Figure 3

Chapter 1

General introduction to hornets and yellow-jackets

Hornets and wasps; how are they related and what is the difference?

There are over 900 species of social wasp (Vespidae) that inhabit all the major land masses, except Antarctica. All social wasps fold their wings down the length of their body at rest, their eyes are kidney shaped and their mid leg has 2 (not 1) spurs. Social wasps have been around for a long time (40-50 million years) with the first ones found in Baltic amber. There are two large families, the mainly tropical Polistinae (paper wasps) that includes by far the most diverse group of social wasps, with many swarm-founding species, bizarre nest structures (Figure 3) and colonies that can contain over a million adults.

Figure 3. A selection of social wasp nests showing the wide range of nest structures. These include nests of Epipona, Synoeca, Polistes, Protopolybia, Polybia, Apoica wasps, and the massive nest of Agelaia that can contain over 1 million wasps. Photographs by S. Martin.

There is even one group of wasps known as 'honey wasp' that produce and store honey. In Brazil, we collected a colony that contained over 20,000 workers and hundreds of queens, but I can confirm the honey tastes good and the onlookers can confirm that the wasps sting is very painful, despite their small size. The other big family are the mainly temperate Vespinae that contains the hornets (*Vespa*), and yellow-jackets, which belong to either the *Vespula* group (build large nest in cavities) or the *Dolichovespula* group (build smaller nests in trees and bushes) (Figure 4). The use of common names is the cause of many problems between scientists and the general public. The term 'wasp' is used by the general public to describe any 'not-hairy' narrow stinging insect, while bee is used for more rounded and hairy insects. However, social wasps, or wasps for short, refer more narrowly to the wasps in Europe that build paper nests, in the USA these social wasps are called yellow-jackets. So, through-out this text I have tried to use the narrower and more accurate 'yellow-jacket' to encompass the social wasps.

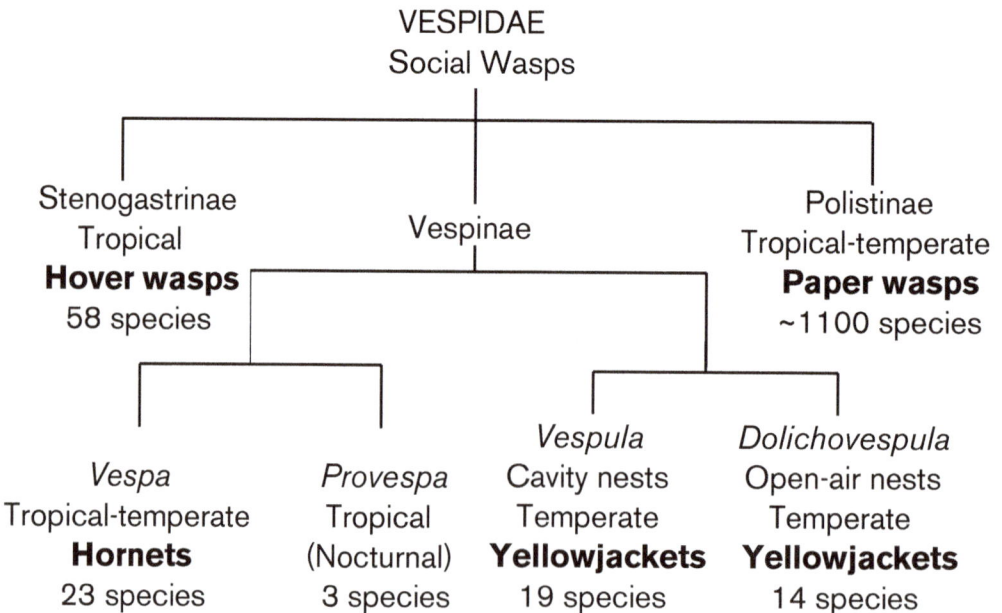

VESPIDAE
Social Wasps

Stenogastrinae	Vespinae	Polistinae
Tropical		Tropical-temperate
Hover wasps		**Paper wasps**
58 species		~1100 species

Vespa	*Provespa*	*Vespula*	*Dolichovespula*
		Cavity nests	Open-air nests
Tropical-temperate	Tropical	Temperate	Temperate
Hornets	(Nocturnal)	**Yellowjackets**	**Yellowjackets**
23 species	3 species	19 species	14 species

Figure 4. An overview of the different social wasp families, how they are related to each other and the number of described species in each family. The common names for each group is written in bold.

Figure 5. Top and underneath views of the Asian hornet that is a typical hornet with a robust body with the abdomen (gaster) connected to the thorax via a narrow petiole, which allows all wasps to curl their abdomen underneath themselves and sting threats. They have two pairs of wings but the hind and fore wings are connected so well via a row of hooks that they appear as a single pair of wings. Photographs by Y. Sakai.

The way to distinguish yellow-jackets from hornets is simple. Hornets are big robust insects, normally over 2cm in length and everything smaller than this is a wasp (Figure 5). However, the correct way to determine hornets from yellow-jackets is via the thickness of the head behind the eyes. So, when looking directly down onto the top of the head see if the distance from the compound eye to the back of the head is greater than the distance from the outside ocelli to the compound eye, if it is greater it's a hornet and if not it's a yellow-jacket (Figure 6). Obviously, this can only be done with dead individuals or a photograph taken at the correct angle. The only other species often mistaken for hornets are large hoverflies, but they only have one pair of wings that are always held away from the body at rest and horntails, which have a very obvious tail that looks like a sting (see Figure 12).

Figure 6. How to identify wasps (yellow-jackets) on the left and hornets on the right, which relies on the distance between the back of the head and the ocelli verses the distance between the ocelli (the three small eyes on top of the head) and inner edge of the large kidney shaped compound eye.

The scientific name *Vespa* is Latin for wasp coming from *Vespillo* that means 'the undertaker whose duty is to carry off the corpses of the poor for burial at night'. Originally wasps were believed to breed in dead carrion, as adults were seen emerging from dead animals, but it was later realised they were collecting food. In hornets and yellow-jackets the queen is normally easy to distinguish since she is the largest individual in the colony. The males are more difficult to distinguish from workers, but the males' antenna consists of 13 not 12 segments as found in females (queens & workers). To the trained eye this addition segment makes the antenna appear noticeably longer (Figure 7). Furthermore, no males of any wasps or bees have a stinger, since the sting evolved from an egg laying tube. Although, if picked up the males often curl their abdomen and try and sting you, which causes most people to drop them!

Female antenna
12 short segments

Male antenna
13 long segments

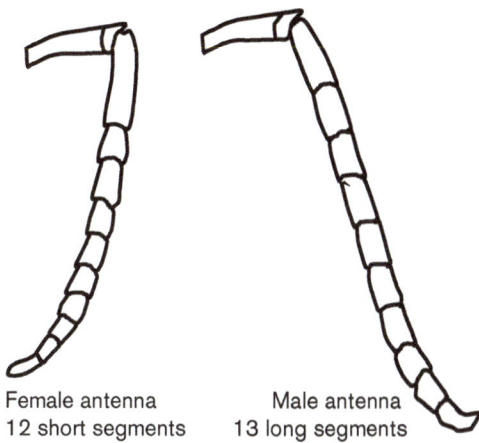

Figure 7. Comparison of the shorter female and longer male hornet antenna's (left) that gives the males a distinctive backward sweeping long antenna (right). Photograph by C. Luck.

The basic structure of the hornet is similar to a honey bee, with a head, thorax and abdomen. Vison is very important and along with the large kidney shaped eyes are the three ocelli on top of their head. These ocelli are enlarged in the single nocturnal hornet species and in all three species of the *Provespa*, a family of entirely nocturnal species found in the Asian tropics.

Hornets and yellow-jackets of the world

Around the world, according to Michael Archer, a world expert in wasp taxonomy there are 22 species of hornets (*Vespa*), 23 species of *Vespula* and 19 species of *Dolichovespula*, which together are called yellow-jackets (Figure 4). The highest diversity of both hornet and yellow-jacket species is found in the highlands of western and central China (Figure 8). This is believed to be their evolutionary birth-place. From there the hornets have successfully spread out throughout Asia and right across the Northern Hemisphere. Going east from China the number of species drops to six in Japan, whereas only two species have spread as far as Europe. Hornets, unlike the yellow-jackets never reached the New World. This is mainly due to their size, since big insects such as hornets need more food and take longer to produce the next

Hornets

Yellowjackets

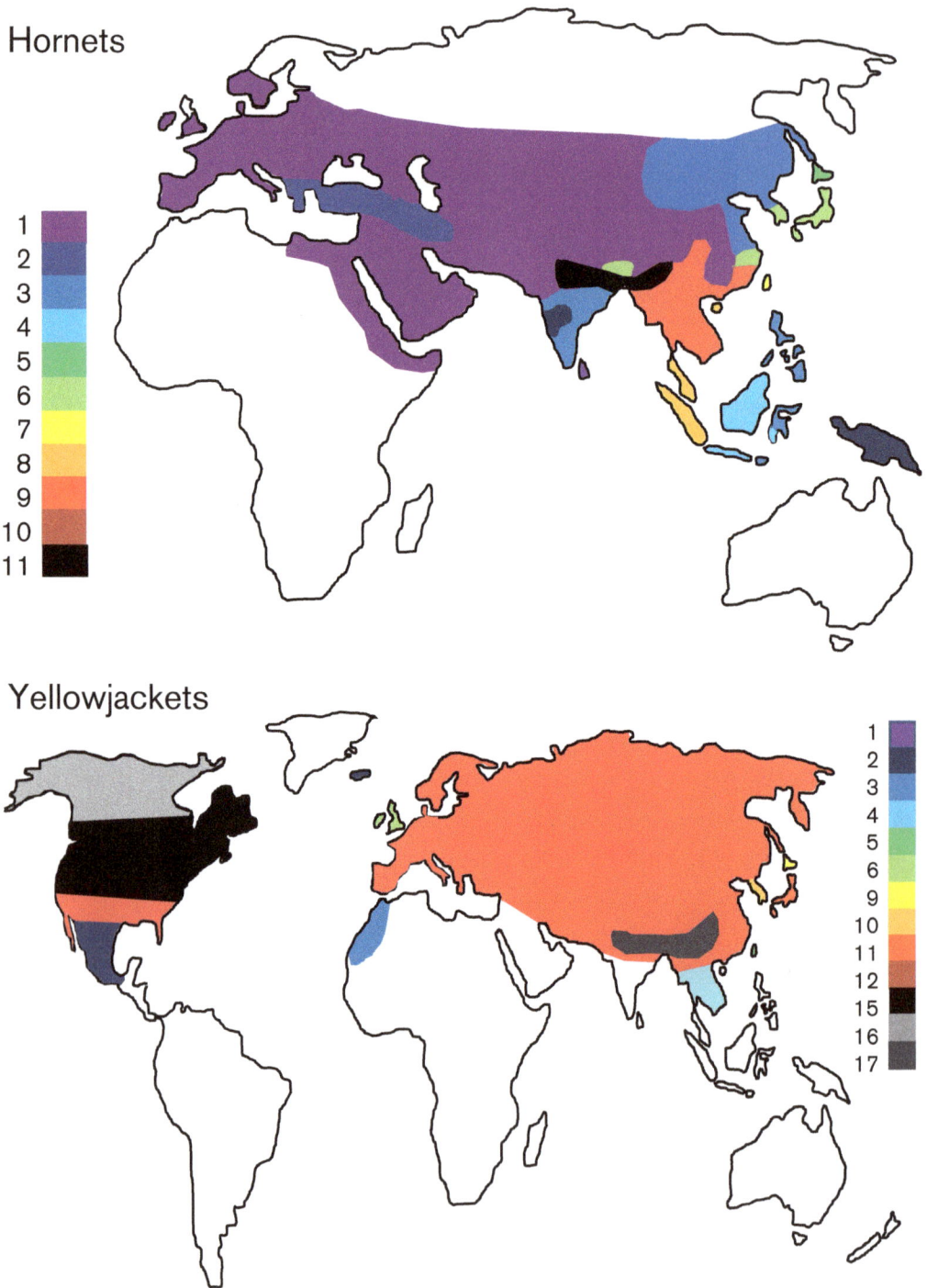

Figure 8. The natural world distribution and species density of hornets (upper map) and yellow-jackets (lower map). The different colours indicate the number of species in that region. (Data compiled from Archer 2012)

batch of sexuals, whereas the smaller yellow-jackets are better adapted to colder climates (Figure 8) and have faster reproductive cycles. This ability to persist in colder climates allowed yellow-jackets to cross the Bering land bridge and enter the Americas where, free from competition, they were able to diversify to become a biological hotspot for yellow-jackets (Figure 8). In tropical climates, yellow-jackets are probably outcompeted by the very abundant Polistinae paper wasps.

Hornets and yellow-jackets of Europe

In the UK and throughout most of Europe we have only one native species of hornet (*Vespa crabro*), and eight species of yellow-jackets (Table 1), although the Saxon yellow-jacket arrived in the UK in the late 1980's, followed by the Media yellow-jacket in 2016. In this case both species arrived from Europe as their natural ranges expanded, possible due to climate change, which is also allowing the hornet to continue to spread northwards in the UK. It can now be found north of the Humber river, although its strong-hold remains along the southern counties (Figure 9). However, *Vespa crabro* is the most widely distributed hornet in the world occurring from the UK in the west right across

Figure 9. The approximate distribution in the UK (top left) of the European native hornet Vespa crabro *(top right) and for comparison the Oriental hornet (bottom), which is found only in Southern Europe e.g. Sicily, Malta, Albania & Greece. Photographs by J. Phipps.*

the Palearctic to Japan in the east. It consists of a species complex composed of eight sub-species, based on different colour variations, with *Vespa crabro gribodoi* being the sub-species (colour form) present in the UK. (Figure 9)

Vespa crabro (Hornet; European Hornet)
Vespula vulgaris (Common wasp)
Vespula germanica (German wasp)
Vespula rufa (Red wasp)
Vespula austriaca (The parasitic cuckoo wasp)
Dolichovespula norwegica (Norwegian wasp)
Dolichovespula sylvestris (Tree wasp)
Dolichovespula saxonica (The Saxon Wasp)
Dolichovespula media (The Media Wasp)

Table 1. The scientific and common names of the yellow-jackets (social-wasps) and hornets found in the UK and throughout most of Europe.

If you live in or travel to more southern areas of Europe you may encounter the Oriental hornet (*Vespa orientalis*), very distinctive and easy to recognise, as it has a reddish body and yellow tail (Figure 9). This species was used as a symbol for lower Egypt by the Pharaohs, although many of the wasps from the shape of their abdomen indicates they were in fact paper wasps belonging to the *Polistes* group.

Chapter 2

Hornet and yellow-jacket invasions

Accidental introductions of wasps and hornets around the world

The accidental introduction of yellow-jackets and hornets to areas outside their natural range is a much more common phenomenon than people realise. Spread has always been via accidental movement of mated hibernating queens, since unlike honey bees, mature nests are not normally 'accidentally' moved. Potential routes include queens overwintering in building materials, Christmas trees or horticultural pots just to name a few. With an ever-con-

Hornet species	Failed introductions	Successful introductions
Oriental hornet *Vespa orientalis*	Madagascar, Mexico, Fujian province of China, Belgium, UK	
Yellow hornet *Vespa simillima*	British Columbia, Canada	
Vespa affinis	California and New Zealand	
Vespa crabro		USA & Guatemala
Asian hornet *Vespa velutina*		France & South Korea

Table 2. A list of hornet species known to be accidentally introduced into foreign countries and whether the introduction failed of became established.

nected world with ever-increasing global movement of goods, commodities and people this has resulted in various species being introduced into many new countries, most introductions fail to establish a new viable population, and these normally go unnoticed but a small proportion do get recorded (Table 2).

Despite several known accidental introductions of various hornet species into a range of countries none persisted except for the two Asian hornet introductions and a single introduction of *Vespa crabro* into North America via New York in around 1840 (Figure 10). Over the next 150 years the distribution of *Vespa crabro* spread west to the Dakotas, south to Florida and Louisiana, and more recently north into the Canadian states of Ontario and Quebec. From the USA, it then spread to Guatemala. Unlike the hornets, the yellow-jackets have established major populations in several oceanic islands including Hawaii, Accession Island, the Canaries, as well as in New Zealand, southern Australia, Tasmania, Chile and South Africa (Figure 10). In all cases, these populations became well established and are very damaging to the local biodiversity, especially on islands where the yellow-jackets outcompete many local species, even birds (honeydew specialists in New Zealand). This is because social insects (ants, termites, wasps and bees) are among some of the most successful invasive groups on our planet, owing to their high reproductive rates, high dispersal abilities, general diet, behavioural flexibility and superior competitive abilities.

What traits helps wasps to invade

So, what are the unique combination of traits that allow yellow-jackets and hornets to get established outside their range? There appear to be several key factors; 1) the ability to establish an entire colony from a single mated female, 2) a hibernation period during which time accidental transportation can occur and 3) polyandry, which is the behaviour of mating with several males.

Although mating with several males is common in the honey bees, this represents the extreme end of the spectrum, since the vast majority of social insect species (ants, termites, bees and wasps) just mate with a single male. For example, the average mating frequency for most hornets is just above one, since the vast majority of queen's mate with a single male. A low number

Figure 10. Accidental introductions of the Asian hornet (red) and European hornet (yellow) and yellow-jackets (green) that successfully established viable thriving populations.

mate with two or even three males, but these are rare events. So, the average mating frequency is just above one. This may help explain why many species of hornets and yellow-jackets fail to establish a viable population when accidentally introduced in a new country.

However, there are exceptions to this general rule, such as the Asian hornet and the several species of yellow-jackets, (*Vespula vulgaris*, *Vespula germanica* and *Vespula pensylvanica*), which are major pests in New Zealand, South America and Hawaii respectively. All these species have an average mating frequency of two or above with some queens being mated by up to seven males. So, in the case of *Vespa crabro* getting established in the USA, a species which normally mates with a single male, but mating with up to three males has been detected. So, maybe only promiscuous queens have any chance establishing a new viable population, but all this is currently just an interesting association, until proven one way or the other.

It appears counter-intuitive that a single hibernating queen that has been fertilized by a small number of males, transported to a distant land can start an entirely new population. Since a genetic bottle neck will occur, as only the genetic material of a single female and a few males is available to start an entirely new population (Figure 11). This results in a massive loss of genetic diversity and loss of vitality, which normally causes the population to fail, but we know this does not always happen. The part of the answer may lie in the study by the great bee scientist Tom Seeley and his team, who were studying a wild honey bee population in the Arnot forest. His honey bee population had gone through a very tight bottleneck caused by the Varroa mite killing almost all of the colonies, but somehow all the colonies did not die and made a steady recovery from a very small number, possibly even one colony. This is known since all the bees in the forest now have the same DNA in their mito-chondria (Figure 11). Mitochondria are only passed onto the next generation through the female line, so it is a way of identifying that a population has gone through a genetic bottle-neck. However, the important nuclear DNA gene diversity derived from the queen and several males was quickly restored to levels seen before the bottle-neck. This is due to gene recombination (gene shuffling) and maintenance of genetic diversity via multiple mating.

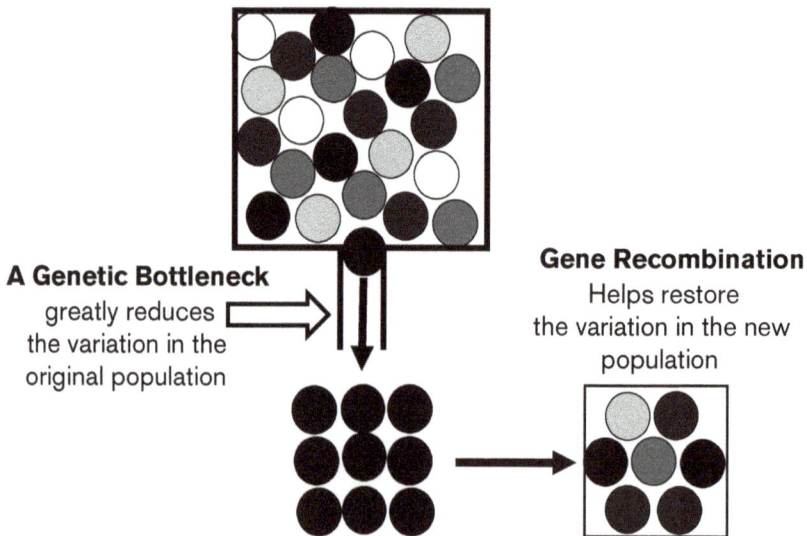

A Genetic Bottleneck
greatly reduces
the variation in the
original population

Gene Recombination
Helps restore
the variation in the new
population

Figure 11. Illustration of the way genetic variation is reduced after passing through a genetic bottleneck. That is initially the population has several variations of a genetic trait, however, after passing through a bottleneck, such as one caused by a single hibernating queen being introduced into a new country, only one of the traits survives, which is shared by all members of the new population. However, then a high gene recombination rate found in Hymenoptera can then restore much of the lost variation.

The Asian hornet and its European cousins

The key distinguishing feature of the Asian hornet is that it is the only hornet in Europe that has distinctive yellow legs, hence its more accurate, but longer and less exotic, common name of the 'Yellow Legged Hornet'. Additional features are largely the black bodies, rather than reddish or brown colour and its size, which is smaller than the native European hornet, but larger than any of our yellow-jackets. According to the excellent Centre for Ecology & Hydrology 'Asian hornet watch' app, the three most common species mistaken for the Asian hornet are the European hornet, the wood wasp or horntail (*Urocerus gigas*) with its very long egg laying tube and the giant or belted hover-fly (*Volucella zonaria*) (Figure 12).

Figure 12. Examples of other species often mistaken for the Asian hornet, such as the wood wasp or banded hornetail (Urocerus gigas), *Giant hoverfly* (Volucella zonaria) *and the Native hornet* (Vespa crabro). *Photographs by G. Barker (wood wasp); A. Gómez (Hoverfly) and J. Phipps (hornet).*

These mistakes are made because many species of insects (beetles, moths, even pupa) mimic the yellow and black markings of wasps and hornets to prevent predation by birds, frogs etc.

Some species like the European and Asian hornets have very large distribution, and so can consist of a species complex, within which many distinctly different colour forms exist. The Asian hornet consists of a species-complex made up of 12 distinct sub-species based on variations of their colours. The sub-species known as *V. velutina nigrithorax* was the one that invaded both South Korea and France. Eight colour forms are restricted to islands mainly found in Indonesia and four colour forms are found on the Asian mainland (Figure 13). The existence of these different and distinctive colour forms provides entomologists with vital information on where an invading species came from and more

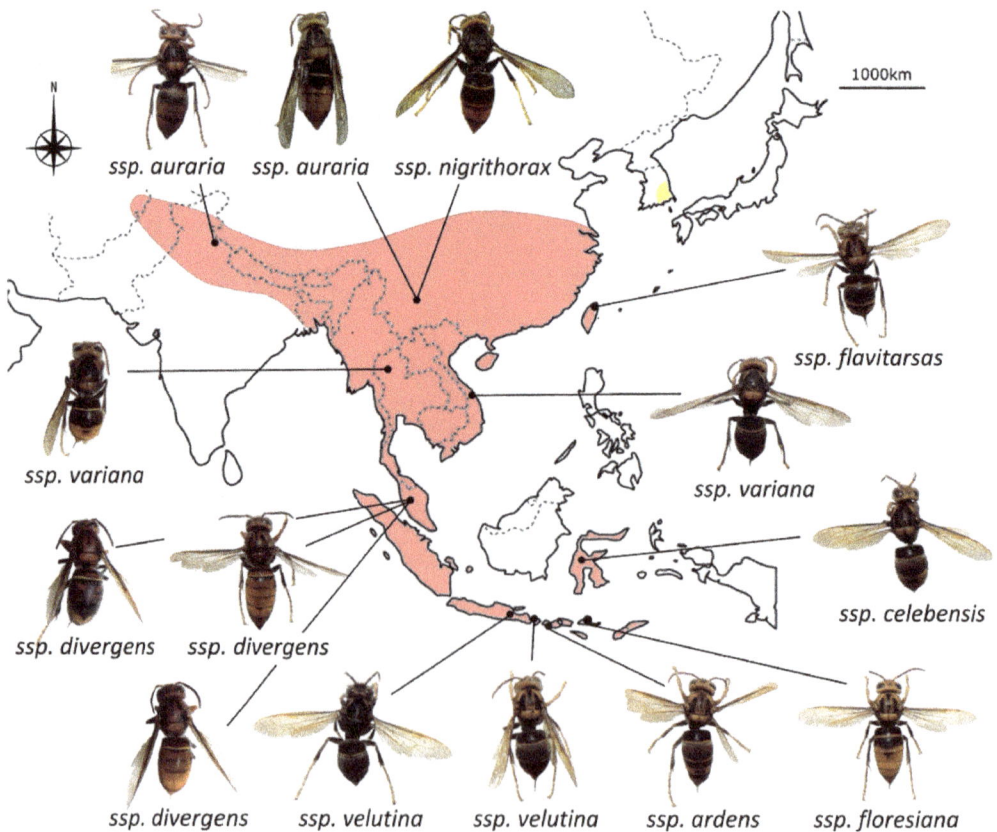

Figure 13. A map courtesy of J. Takahashi showing the location of ten of the known twelve sub-species of the Asian hornet. The sub-species that invaded France and South Korea is Vespa velutina nigrithorax *found in central and Eastern China.*

importantly, what environmental conditions it is adapted to. For example, the Asian hornet has evolved into a wide range of sub-species, each with its own habitat preferences, some are adapted to survive in the hot lowland tropical plains of Indonesia, whereas others are adapted to high mountainous areas in the Himalayas, living between 3,000-10,000ft. Therefore, the accidental intro-duction of a sub-species adapted to survive on a hot tropical island may not pose the same initial threat as a sub-species coming from the mountainous central areas of China, which has a better chance to thrive in the temperate regions of Europe and Korea. As always in biology, the devil is in the detail.

The Asian hornet *(Vespa velutina)* arrival in France and subsequent spread

During the winter of 2005 two large spherical hornet nests appeared high up in trees in the Bordeaux region of France. These were destroyed by the local farmer using a shotgun, although they would have already produced their

Figure 14a. The map shows the actual spread of the Asian hornet across France over the past decade (for more details see paper by Barbet-Massin et al., 2013).

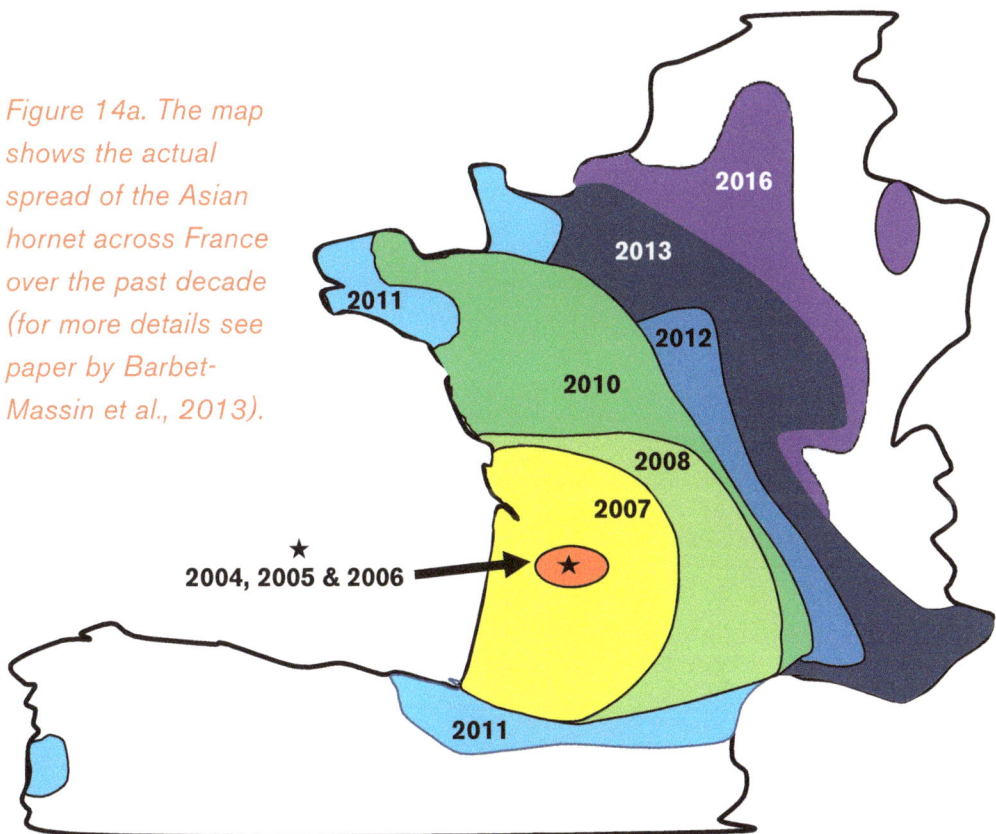

2016

2013

2011

2012

2010

2008

2007

★
2004, 2005 & 2006

2011

Figure 14b. The map indicates the predicted distribution of the hornet under the current climatic conditions (for more details see paper by Barbet-Massin et al., 2013).

sexuals by the time they were shot. Later it emerged that three nests had been seen in 2004 in the same region. Therefore, it is likely that a small number or even one, hibernating queen arrived during the autumn/winter period of 2003/2004, or even earlier. Since several surveys showed the Asian hornet quickly spread out across France from this single focal point near Bordeaux (Figure 14a), the rate of spread has been estimated as an amazing rate of 78km per year. So, it soon became widespread throughout most of southern France, and became established in the Basque Country in Northern Spain. It appeared in Portugal and Belgium and eventually reached Italy, Switzerland, Germany with sightings in the Channel Islands and the UK (see below).

The hornet's spread was typical of many invasive species, that is natural local dispersal aided by human assisted accidental long-distance movements. Although the role of possible migration (see hibernation) cannot be ruled out, especially in light of the rapid local expansion. So, in just 12 years since the introduction of just one or two hibernating queens, the Asian hornet has established a very large and flourishing population right across France, which is now spilling over into the surrounding countries.

Determining the final limits of the Asian hornets' new distribution is very diffi-cult. However, using a previous model of environmental similarity and today's known distribution in Asia, the Asian hornet is predicted by a team of French scientists to continue to expand its range for some time (Figure 14b), although there is a good chance it will eventually extend even further across most of northern Europe as far as the 60°-degree latitude line, beyond which hornets cannot persist (Figure 8).

The Asian hornet arrives in the UK

The density of hornets increased across France and has reached up to ten nests per square kilometer in urban environments. This occurred in just seven years after it was first detected and despite a policy of local colony destruc-tion in France. With such high numbers of hornet queens being produced in France, it was only a matter of time until a hibernating queen turned up in the UK. The hornets' arrival was preceded by numerous false and often humorous reports by the media, which included a 'three inches long' deadly Asian hornet with the sting as big as a needle that turned out to be a harmless wood-wasp (Figure 12); A swarm of deadly Asian hornets that attacked run-ners in the grounds of Downton Abbey and various native hornet sightings. The first actual sighting was reported from the Channel Island of Alderney where a small nest containing around a dozen workers and one large queen in a storage shed was found and removed in the early summer of 2016.

Then in mid-September of 2016 the first confirmed mainland sightings were reported by a beekeeper in the Tetbury area of Gloucester in the South-West of England, many miles from any major port. This was predictable since the accidental transportation of a hibernating queen can turn up anywhere, due to the massive amount of movement of people and goods by cars and lorries between France and the UK. In Tetbury almost two weeks after workers were reported on the 30th September 2016 a nest was located and destroyed. The nest was found at the top of a 55-foot-tall conifer tree, which is very typical for this species (Figure 15).

Subsequently on the 4th October another hornet was reported over 40 miles away in the Mendip Hills in Somerset. This individual was of unknown sex or caste, since it was already desiccated when reported, so its true origin

Figure 15. Typical Asian hornet colonies high up near the top of large trees, although colonies have also been found on cliffs. Photographs by T. Yamamura.

remains a mystery. A subsequent search of the area failed to locate a colony, if one was indeed present. Then on 17th March 2017 another Asian hornet (presumably a hibernating queen) was found in a retail warehouse in central Scotland, indicating again the random nature of accidental introductions. So, the Asian hornet has reached the UK and accidental importations of this species will continue long after the hornet is well established in the UK. The

The Register
Biting the hand that feeds IT

Science   21 Sep 2016

Asian hornets are HERE... those honey bee murdering BAS*DS
Flying death spotted in the Cotswolds, boffins move in for the kill**

Figure 16. One of the more humorous media reports that announced the arrival of the Asian hornet to the UK.

arrival of the Asian hornet in the UK was accompanied by some over-the-top reporting by the media (Figure 16).

Was the Asian hornet eradicated from the UK?

Due to the size and public awareness of the Asian hornet, it is likely that one or more mated hibernating queens arrived in the UK during the autumn/ winter period of 2015/16. Based on data from the yellow hornet (*Vespa simil-lima*) a species with a very similar biology, the first batch of queens emerged around 147 days after the start of the nest, which is estimated as May 6th. So on the 30th September, the day the UK colony was destroyed, it is likely to have contained new queens and males. Although, it is impossible to predict with any certainty if any sexuals had already left the colony prior to it being destroyed as newly emerged queens and males spend several days within their nests prior to leaving. However, even if all Asian hornets were eradicated in 2016 the continuing threat from the accidental importation of queens from France will always remain.

The Asian hornets' arrival in South Korea and subsequent spread.

On the other side of world an almost parallel story was developing, since the same sub-species of the Asian hornet that invaded France had also been accidentally introduced into South Korea. The first sightings were in 2003 around the southern port city of Busan. Unlike Europe, South Korea is already home to six species of hornets so its initial arrival could have been easily

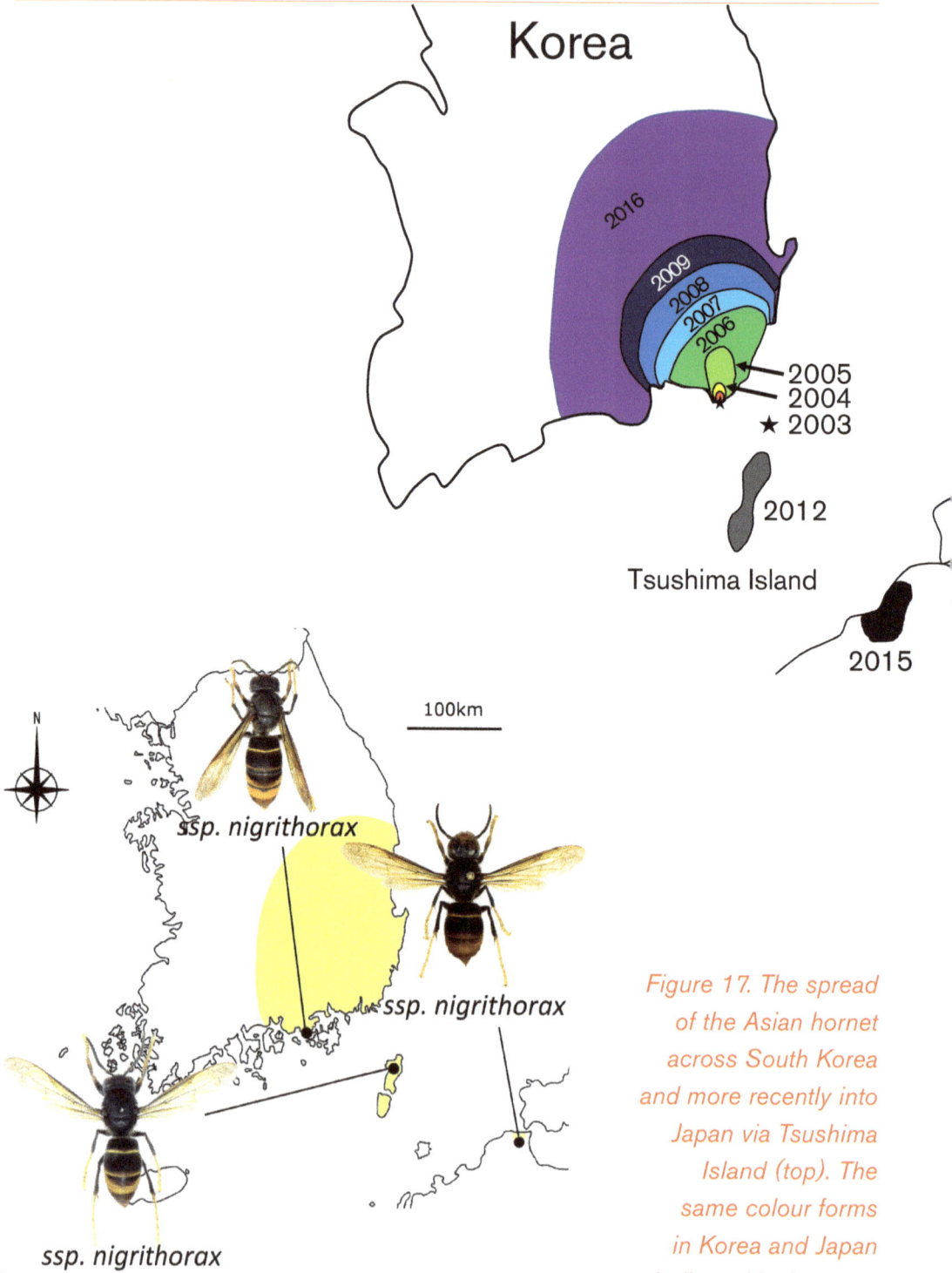

Figure 17. The spread of the Asian hornet across South Korea and more recently into Japan via Tsushima Island (top). The same colour forms in Korea and Japan indicate it's the same population (bottom). Image courtesy of J. Takahashi.

missed. The Asian hornet quickly became established and started to spread northwards radiating out from Busan across South Korea (Figure 17). This population slowly expanded its range by around 30 km per year, so by 2013 the hornets' distribution covered the southern half the South Korea. Subsequent genetic analyses by scientists from both France and Japan confirmed that the South Korean population originated from China, just like the French population. No diversity in a mitochondrial gene, both in the South Korea and French populations, indicates that both populations originated from a single or a very small number of founder queens from the same region of Eastern China (Zhejiang or Jiangxi districts). It is possible that in Eastern China during 2001/2002, the Asian hornet population in Eastern China experienced a 'wasp-year', were the local conditions allow large number of colonies to be successful, resulting in a large number of mated queens being produced and so increasing the chances of them been accidentally exported.

As densities of the Asian hornet increased in South Korea, just as was happening in France, so does the likelihood of them being transported further afield. So, in 2012 nests of the Asian hornet were sighted on the small Japanese island of Tsushima. This island lies 50km from Busan in South Korea and 50km from the mainland of Japan. Although on a clear day Tsushima Island can be seen from the mountains of South Korea it is unlikely the queens would be able to reach it under their own steam. However, Tsushima Island is a popular destination for both Korean and Japanese tourists, which provided the hornet with a stepping stone to mainland Japan, which it reached in 2015, when it was reported from Northern Kyushu (Figure 17).

Chapter 3

Hornet life cycle

There are two main ways for social wasp colonies to reproduce. Firstly, a single queen establishes a new colony every year often after a period of hibernation, this includes all hornets, even in the tropics. The other way is a new colony can 'bud off' from the main colony, as found in honey bees and all swarm-founding wasps. This is when part of the existing colony swarms, containing some of the queens and workers. Swarm-founding wasp species are usually associated with the tropical regions where food (insects) are present throughout the year. As honey bees have evolved a food storage system (honey) this allows them to survive periods when food is absent (winter), but this is not an option for carnivorous insects like wasps living in temperate environments.

As life cycles are a continuously connected circle, it is not possible to define the start or end point, so I have chosen to start during the hibernation phase, since this is where the colony size is smallest at just one individual - the mated queen. Furthermore, until sufficient data is available, I have used data from the yellow hornet (*Vespa simillima*) in place of the Asian hornet, since these two species have almost the same life history, adult size and colony growth patterns.

Hibernation (Oct/Nov to April)

In all hornets, yellow-jackets, bumblebees and many other species of social insects, each colony is started by the efforts of a single queen. She was mated the previous year and has spent the winter hibernating in a place pro-

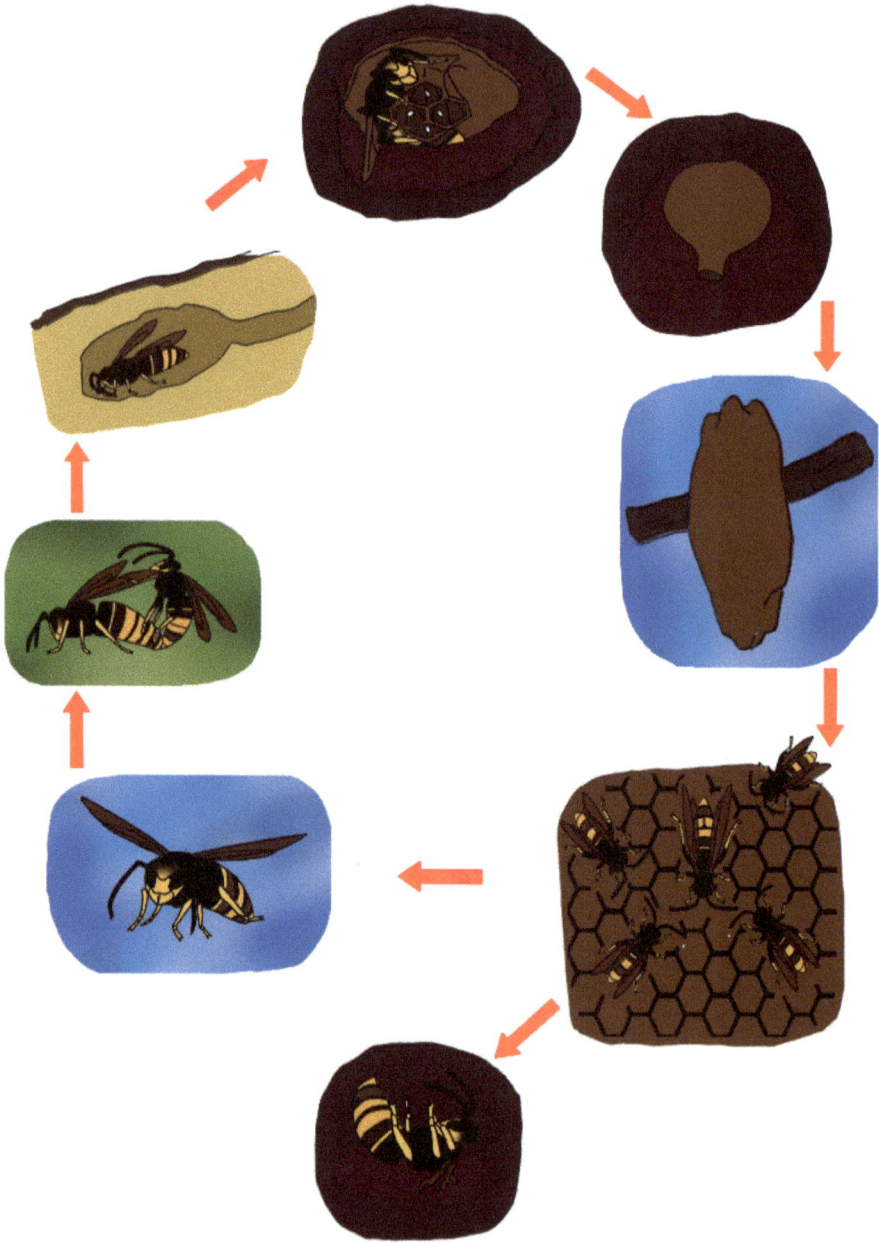

Figure 18. General life cycle of the Asian hornet. Starting at the top, the queen emerges from hibernation and builds an embryo nest. This develops until it fills the cavity then relocates into a tree. Here the colony grows to produce many workers that live for a few weeks before dying. Finally, sexuals (males and queens) are produced. These leave the colony and mate. The queen spends the winter hibernating while all the other hornets die. Image designed by J. Takahashi.

Figure 19. A hibernating hornet queen in the typical position with wings tucked beneath her abdomen. Photograph by S. Martin.

tected from the rain, snow and wind, such as in a pile of old carpets in a shed, under the roof of a bee hive, or in a rotten log. During hibernation queens tuck their wings under their abdomen pressed close to their body, giving them a very distinctive pose (Figure 19).

Rising air temperatures during spring awaken the queen from her deep winter sleep. Often the activities of humans can bring queens out of their hibernation prematurely, for example by turning on the heating in a shed, or holiday home. Also, if a particularly warm spell of weather occurs in late winter, this can cause the queens to come out of hibernation, so wasp and bumble-bee queens can often be seen flying around. This is an important impact that climatic change can have, since once broken, re-entering hibernation is very difficult if not impossible and the queens die of starvation. Almost all (97%) hibernating queens are found to be mated, therefore, any overwintering queen is capable of established a new viable colony.

In Japan, based on a large number of observations between 1966-1975, Makoto Matsuura, a master of hornet and wasp research from Japan, found the average appearance of the yellow hornet queens occurred on 16th April, while the European hornet appeared over two weeks later on 2nd May. Therefore, you can predict that the Asian hornet queens would start appearing in mid-April. As soon as the queen emerges from hibernation she starts to seek nutrients, since all her fat stores are exhausted as they have been used up during the past six months (Figure 20). The queen feeds on any nectar or tree resin available (Figure 21), which helps activate her ovaries and sustain her food foraging activities and nest-building, since no new fat is laid down.

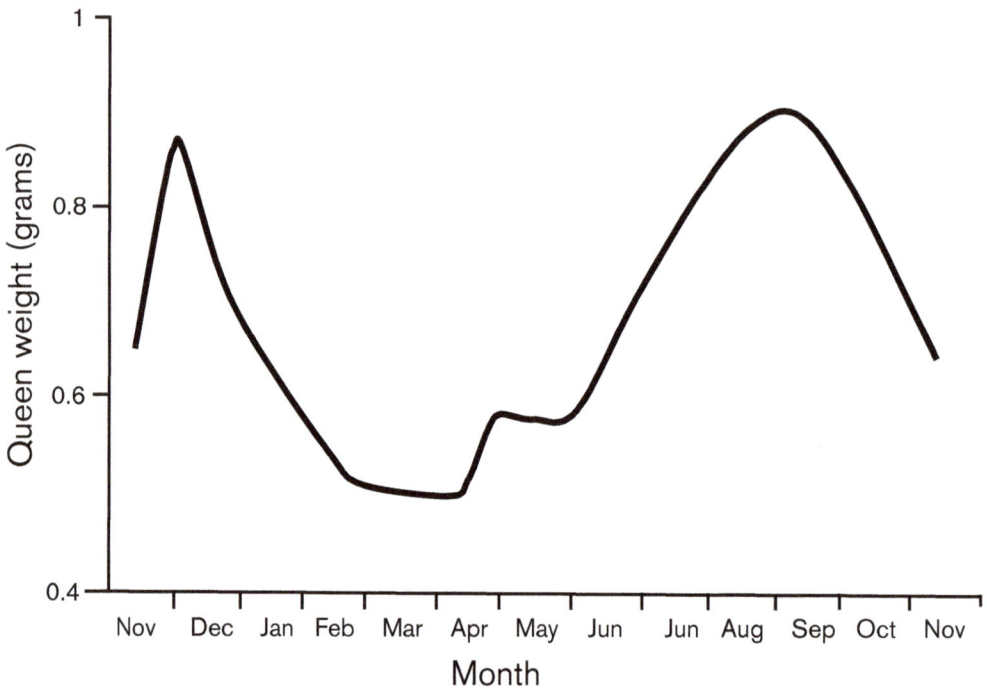

Figure 20. The changing weight of a hornet queen over the year. She is heaviest in late autumn due to stored fat and again in the following September due to her enlarged ovaries. In spring, the queen is at her lowest weight as she emerges from hibernation, quickly gaining weight from nectar and tree resin, before stabilising during the embryo nesting phase, then increasing from June onwards as her ovaries enlarge due to increased egg laying. (Adapted from Martin, 1992).

Post-hibernation migrations

Although not often recognised, post-hibernating queens of several species of hornets and yellow-jackets have been recorded or have been assumed to make migration flights prior to starting a colony. Migration has been recorded in the European hornet, whereas, in the Giant Japanese and yellow hornet, hundreds of marked queens remained in the study area for one week feeding on tree resin (Figure 19), then completely disappeared from the study area and the surrounding vicinity, indicating that migration may have occurred.

Furthermore, in a flight mill an Asian hornet queen was able to fly an amazing distance of 200km, but this is unlikely to be achieved under natural conditions. However, the evidence for post-hibernation migration in yellow-jacket queens is much stronger. In some years, many thousands of *Vespula rufa* and *Dolichovespula saxonica* queens per hour have been seen moving South-West along the Finnish shoreline in mid-June. This migration behaviour, although not yet recorded in the Asian hornet, may help explain the rapid expansion throughout France, since it is an effective long distance, natural dispersal mechanism. Insect migrations are notoriously difficult to observe unless they occur in the form of a mass migration containing thousands of insects all heading in the same direction, which is very unlikely to occur in hornets until their densities saturate the environment.

Figure 21. Queen hornets feeding on tree resin or nectar to gain nutrition early in the season. Top left (Vespa tropica); *Bottom left* (Vespa affinis); *Top right and bottom right are the Asian hornet. Photographs by S. Martin (Left images), and Y. Sakai (Right images).*

The embryo nest (April/May-June), solitary period

During the next few weeks the queen seeks out a suitable nesting site in an enclosed and protected site, such as a wall cavity or garden shed or a hole in a wall or tree. The basic structure of the embryo nest in all hornets is basically the same, consisting of a petiole (stalk), which forms a single point of attachment of the nest to a surface. The lower end forms the initial comb, which contains the hexagonal cells in which the brood are reared (Figure 22). The entire structure is quickly surrounded by an envelope that initially consists of a single layer of paper to which more layers are added later.

Figure 22. An Asian hornet embryo nest with the first few cells containing eggs or young larvae, the envelope has yet to be completed. Photographs by C. Luck.

The entire nest is only 4-5 cm across and is constructed by the queen. She scrapes strips of wood from a dead tree, or other wooden structures such as telegraph poles, fence or shed (Figure 23). The scraped ball of wood pulp is mixed with her saliva and chewed to produce paper which, using her mandibles, she constructs a narrow strip of paper from which the nest is constructed.

The different coloured strips of paper indicate the different sources of wood pulp. During the development of the embryo nest, the queen builds around one new brood cell each day. When completed, the nest entrance of the

Figure 23. Hornet queens of Vespa affinis *(left) and the Asian hornet (right) gathering wood pulp from branches. Photographs by S. Martin* (Vespa affinis) *and Y. Sakai (Asian hornet).*

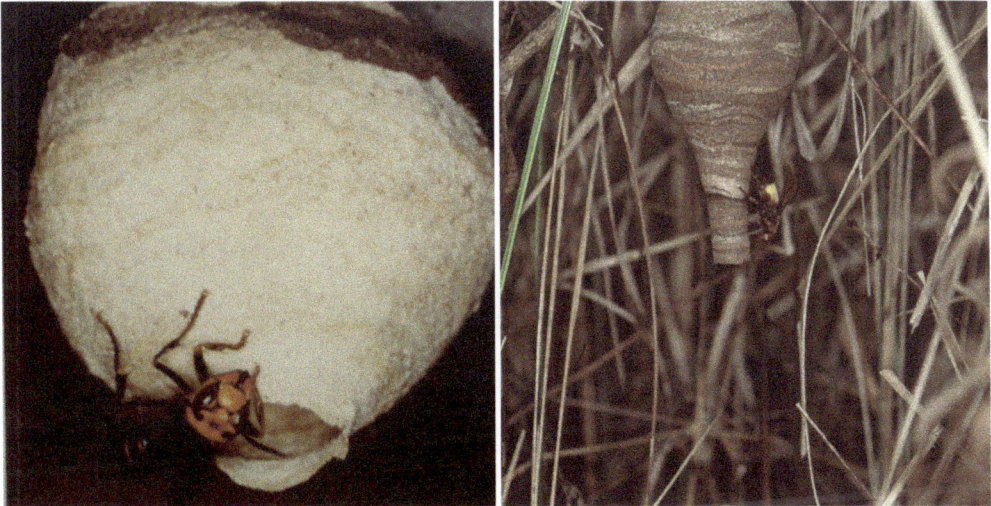

Figure 24. Shows a typical completed hornet embryo colony (left) although a few tropical species build their embryo nest with a long entrance tube (right). Photographs by S. Martin.

embryo nest for most hornets, including the Asian hornet, is a simple hole at the bottom of the nest (Figure 24). Although in other species, like *Vespa analis*

Figure 25. During the night, the queen wraps herself round the top of the comb, in order to help speed up the development of the brood by warming up the comb with the heat of her body. Photograph by S. Martin.

and *Vespa affinis*, they build an elongated entrance tube (Figure 24) although the reason for doing this remains unclear.

Once the first few brood cells are built and the envelope is completed, the queen lays a single egg into the base of each cell, which is attached to the cell wall by an adhesive. After three to four days, the egg hatches into a larva and does not fall out of the cell. The larva's last segment remains firmly attached to the old egg case until it grows large enough to fill the entire cell. Each larva goes through four moults, before spinning a silk cocoon using threads produced by their salivary glands, before moulting a 5th and final time into the pupa stage.

At night, the queen is often found curled around the petiole on top of the comb rather than under the comb (Figure 25), in order to speed up the development of her brood. The queen is warming up the comb and the brood it contains via heat the queen produces by her own body.

Although being a single individual, the rise in temperature is small but still detectable and will help the brood to develop quicker. Several studies on the yellow hornet found it takes around 50 days for the queen to build the first comb that consists of between 35-50 cells in which she will rear the first 10-15 brood into adult workers. The time it takes for each egg to develop into an adult reduces at an almost linear rate as the size of the embryo colony grows. It appears only mated queens attempt to build of embryo nest, since an unfertilised queen could 'in theory' produce an embryo nest that would only contain male offspring, but these types of nests are very rarely, if ever seen.

Protection of the embryo nest from ants

The most vulnerable time for the colony to predators is during the embryo nesting period, since the queen must collect wood pulp and find sufficient food to feed herself as well as her developing brood. So, during this period the queen spends long periods of time away from her nest. The totally defenceless brood make a nice meal for any number of predators especially ants that are expert in finding food within their environment. However, the embryo nest is protected from these marauding ants by an ingenious method. Prior to leaving the nest the queen applies a chemical repellent onto the comb stalk (petiole). This ant repellent is produced in a special set of glands (Van der Vetch glands) found inside her abdomen and is applied to the petiole via an area of short hairs that form a brush on the underside of her abdomen (Figure 26).

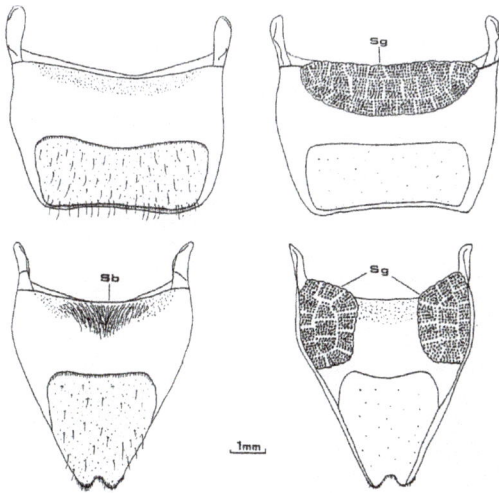

Figure 26. The outside (left) and inside (right) views of the last two segments of a queen hornets' abdomen showing the position of the Van der Vetch glands (Sg) that produce an ant repellent and the sternal brush (sb) that is used to apply the repellent to the comb petiole. Drawn by S. Martin.

Figure 27. The development of the comb of a hornet embryo nest. The top of the comb forms the petiole, which has a dark shiny surface due to the application of ant repellent. Photograph by S. Martin.

This forms a dark surface on the nest petiole that the ants cannot cross (Figure 27). So, although the ants can find the nest they firstly must get inside the nest via the nest entrance, which requires the ants to negotiate a tricky 180° bend. If breached the brood remain protected by the chemical barrier applied to the petiole. The effectiveness of these two lines of defence allows hornet colonies to become established and develop successfully just 30cm from the entrance to an ant colony. The value of this chemical barrier is seen when the queen is artificially removed. Within a couple of the days the ants are feeding on the hornet brood, since the repellent needs to be renewed daily.

Queen usurpation

Despite the constant predation pressure from ants, the hornets' biggest threats come from their own species. If an embryo nest is accidentally destroyed the queen does not rebuild a new nest, but instead tries to invade another nest and attempt a takeover by killing the resident queen. This is known as 'queen usurpation' and it appears to be a common but unseen behaviour that may be a very important population control behaviour in hornets and yellow-jackets. Queen usurpation can result in several dead queens being found in the area around a nest (Fig 28). For example, up to 14 dead queens have been recorded under an active *Vespa affinis* embryo nest.

Although, fights between the queens often lead to injuries, even the victor may be damaged sufficiently to either cause the colony to fail or become smaller than usual. It is possible that any late emerging queen may never attempt to establish their own colony and just attack other ones, which could lead to a population of late emerging 'parasitic' queens. However, this is unlikely as this

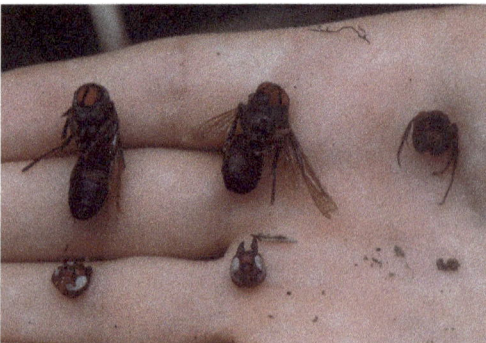

Figure 28. The remains of three different hornet queens found under an embryo nest due to usurpation fights. Within hours of dying the ants remove all the protein from their bodies. Photograph by S. Martin.

would lead to a parasitic live-style with adaptions to flight, which are only seen in the highly specialised 'cuckoo hornet' *Vespa dyboswskii* (see Chapter 4).

Queen usurpation helps explain why the widespread killing of hibernating queens in New Zealand, driven by a bounty paid for each hibernating queen located, resulted in, against all predictions, an increase in wasp colonies the follow year. This is because the vast majority of hornet and yellow-jacket queens fail to successfully establish a new colony and raise the first batch of brood. However, once this is achieved the chances of the colony surviving to produce the next generation of sexuals is high. Removing the number of queens from the population can increase the individual success of the remaining ones. Likewise, in Japan during a year when embryo colonies are abundant and usurpation rates are high, very few large mature colonies appear later in that year. So, queen usurpation may be one key factor that drives the two-year population cycle observed by Michael Archer in the UK yellow-jacket population.

Cooperative period (June)

When the first workers appear, so starts the short co-operative period when both the queen and workers are active outside the colony. Each newly emerging worker chews its own way out of their silk cocoon. These young adults, also called 'callows', are pale in colour, have a soft cuticle (skin) and spend the first one to two days head-down in any empty brood cells, resting. During this time, they occasionally approach their sister larvae to feed (see diet), and their cuticle darkens and hardens. As soon as this period has passed the workers start to work both within and outside the colony, helping the queen in nest building, foraging and brood nursing duties. These first few workers are also the smallest adults produced as that is more efficient. People often comment on the sudden change in the size of bumble-bees and yellow-jackets, since they first see the large queens flying around to be followed by tiny workers. However, within two to three weeks of the appearance of the first workers the queen never leaves the safety of her nest. Now the queen's main role is egg laying, while all other foraging, brood rearing, defence and nest building activities are taken over by the increasing number of new workers. During the co-operative stage, when the colony is small, the workers and

Figure 29. A yellow hornet cooperative nest in an enclosed wall cavity (left). The original pale embryo nest can be seen beneath the more recent darker strips of envelopes. However, within a few weeks the nest will start to expand rapidly as demonstrated by the cooperative nest of the Asian hornet (right). Photographs by S. Martin (Yellow hornet) and C. Luck (Asian hornet).

queen are not that aggressive, since the loss of a single worker will impact on subsequent colony growth. Colony growth is slow during the cooperative period (Figure 29), but as the colony enters the next phase the colony will start to expand rapidly.

The worker nest (June-August) or polyethic period (including nest relocation)

During the co-operative phase, as more workers emerge, there are more individuals to share the work load. This means during the next two to three months both the physical size of the colony and its population increases rapidly. Inside the colony, many changes are starting to occur, such as the workers behaviour towards their queen. During the co-operative period, the queen is treated just like any individual within the colony, but during the pol-yethic period workers start to orientate towards the queen licking her body, which in the later stages develops into a type of royal court, especially in some species of yellow-jackets. This attention by the workers causes the queen to lose all her hairs and become shiny (Figure 30), which can eventually lead to the loss of the queen's wings in a mature colony.

This shiny appearance allows the rapid identification of the egg-laying (founding) queen within a mass of workers when a colony is collected. As the nest grows a series of additional horizontal combs are added. These are joined to the pre-existing combs by a series of pillars. To support the extra weight,

10mm

Figure 30. Upper panel shows the comparison from left to right of a mature 'shiny' egg-laying queen; a newly emerged queen from a mature colony; a worker and male of Vespa affinis *hornets. The large size of the queen relative to the workers and males is obvious in this species. Note the loss of wings and hairs from the egg-laying queen on the far left. The lower panel shows from left to right a queen, worker and male of the Asian hornet that are all much similar in size. Photographs by S. Martin (*Vespa affinis*) and Y. Sakai (Asian hornet).*

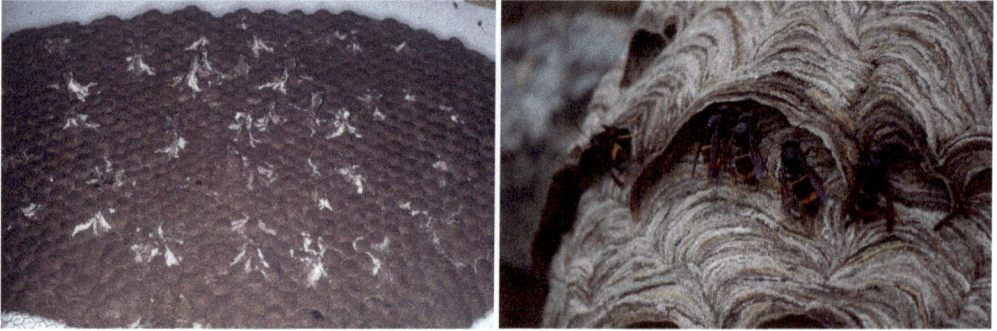

Figure 31. The back of a large hornet comb (left) where the petioles that contact the combs together are strengthened with the silk capping's from the emerged brood cells. While groups of workers continue to build the envelope (right). Photographs by S. Martin (upper) and C. Luck (lower).

the original central petiole is thickened and additional suspensoria (petioles) connect the combs. These are strengthened by embedding old silk pupa cocoons from the emerged brood into the paper (Figure 31), a type of reinforced concrete. The colour of the combs is uniform since the paper used in their formation is taken from the inside of the envelope and re-used to build the comb. So as the workers add new strips of paper to the outside of the nest (Figure 31), other workers are removing the inner surface of the envelope to build the comb, so the entire nest increases in size.

In unrestricted areas, each comb maintains a regular round pattern, since new cells are constructed round the edge of the existing comb. The comb construction pattern is mirrored in the brood pattern with the eggs, larvae and sealed brood, appearing in a series of concentric circles (Figure 32).

Figure 32. An early stage worker nest showing the typical pattern of concentric circles of brood. Photograph by S. Martin.

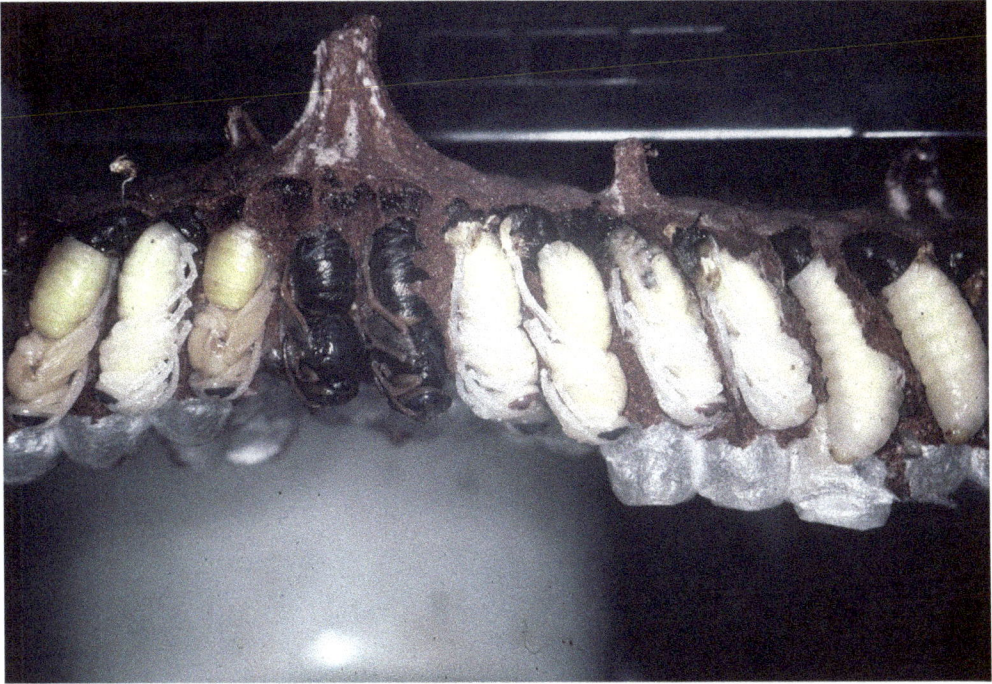

Figure 33. A cross-section through a hornet brood comb containing the development of the pupae. At the base of the cell a large black 'meconium' can be seen. Photograph by S. Martin.

This circular pattern persists as cells can be re-used up to three times in the early combs. That is one cell can produce up to three workers. It is relatively easy to determine the number of times a cell has been successfully used, since when the mature larva pupates, it discards all its undigested faecal material into the base of the cell, where it hardens into a pellet called a 'meconium' (Figure 33). These can easily be separated and counted since each successive meconium is separated by a layer of the next silk pupal cocoon. This can be used to determine how many adults have been produced by the colony.

As the colony grows rapidly, the internal nest temperature starts to stabilise at a steady 30° C (see section on thermoregulation), which along with an increased food supply helps speed up the development of the brood. The length of time to develop from an egg to an adult worker drops from around 50 days in the embryo nest to just 29 days in these large colonies. As the larvae mature they scrape the side of their brood cells with their teeth as a beg-

Figure 34. Vespa dybowskii *(right) and Japanese Giant hornet (left) workers guarding the nest entrance, ensuring nothing gets into the nest, and ready to raise the alarm if any threat appears. Photographs by S. Martin (*Vespa dybowskii*) and M. Ono (Giant hornet).*

ging signal. When they do this in unison it can easily be heard. The entrance is now constantly guarded day and night by a small group of workers (Figure 34), which are now much more aggressive in the defence of their colony.

Colony relocation

One method to increase the chances of nest survival found in the Asian hornet, as well as several other species that build large aerial nests such as the yellow hornet is 'nest relocation'. This is where the queen initially establishes the embryo nest within the protected environment of a natural cavity such as a hole within a tree. Then later builds a second aerial nest where it can expand rapidly in size and allow unrestricted access for the large worker force. So, early in the polyethic period the colony will normally outgrow its original nesting captivity, although this is not always the case (Figure 35).

It is now that the workers usually relocate the entire nest by seeking out a new location, typically for these large nesting aerial species, either high up in a tree or under the eaves of a tall building. This allows the easy access for the increasingly large number of outgoing and incoming foragers from the single entrance, as well as protection from large mammals such as bears, which exist in much of the native range of these species.

During nest relocation, often several starter nests are built and quickly aban-

Figure 35. Unusual yellow hornet (left) and Asian hornet (right) colonies that, rather than relocate, continued to expand within their original cavity, which often forces the envelope to be built of the outside of the structure housing the colony. Photographs by S. Martin (Yellow hornet) and C. Luck (Asian hornet).

doned, before the workers settle on one favoured location and in just a few days built a football size nest (Figure 36). This becomes gradually occupied by the workers and queen over a period of about one month, as the original colony is allowed to run down and the new colony starts to rapidly expand in size.

Up to this point in the nest cycle the combs contain very few cells that do not contain brood since the queen is able to always lay an egg in any newly built or recently vacated cell. However, as the rate of colony growth increases the queen reaches their physiological limit of laying around 100 eggs per day in a species like the yellow hornet. As the colony reaches its maximum size

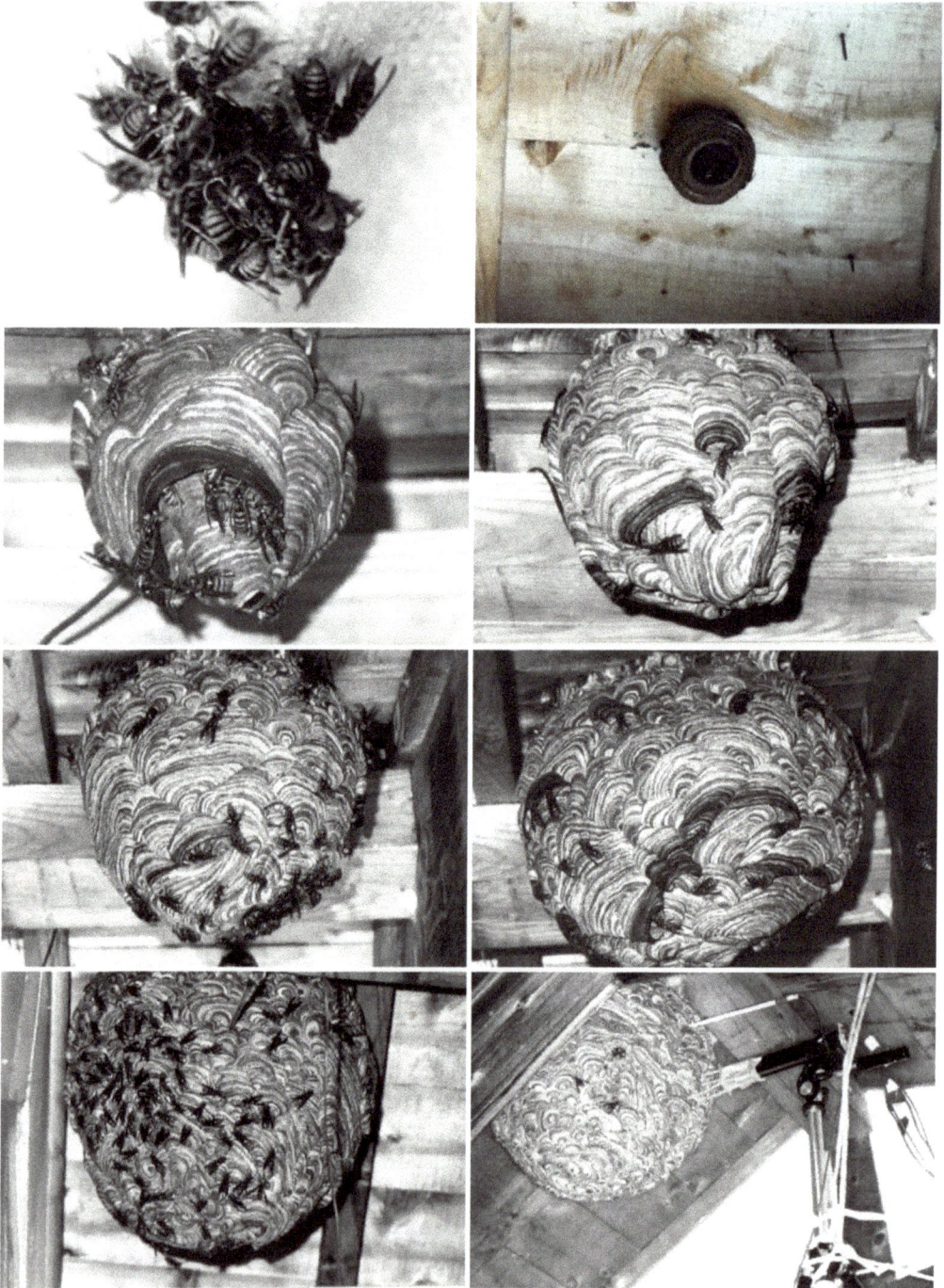

Figure 36. The typical relocation of a hornet colony from the first group of workers that quickly establish the main attachment (top left) and within a few days a new colony is built under an eaves (top right) or in a tree. Over the next few months the size of the colony increases rapidly. Photographs by S. Martin.

Figure 37. Showing the variation in envelope structures between the long sheets of envelope built by the European hornet (left), the small pockets built by the Asian hornet (right). Photographs by C. Luck (left).

low-jackets, so the colony enters the reproductive phase. The structure of the envelope is different in each hornet species (Figure 37) and is useful in determining the species of colony after the colony has died.

Reproductive phase (September-October)

In all social bees, wasps and ants, males are produced by unfertilised eggs and females via fertilised eggs. So, at the start of the reproductive period the queen takes the apparently unusual step of starting to lay male eggs and at the same time reducing the production of worker-destined eggs. Around a week later the queen starts to lay eggs that will be become new queens. The mechanism that determines if an egg develops into a worker or queen remains unknown. As male and queen production increases worker production declines. At some point during this final stage the queen often disappears. The order of worker-male-queen production has a profound effect on the success of colony, associated with the environmental conditions (Figure 38). So, if autumn is cold and wet and the insect populations (hornet food) declines quickly then the colonies will predominantly produce males. However, if the autumn is favourable (Indian summer) or the hornets can find a reliable supply of food (honey bees) then the colony cycle is extended and more queens and males are produced. For example, a colony on 1st Oct can produce 300 queens and 600 males, but by the end of October just 30 days later the same colony can produce 1800 queens and 1800 males!

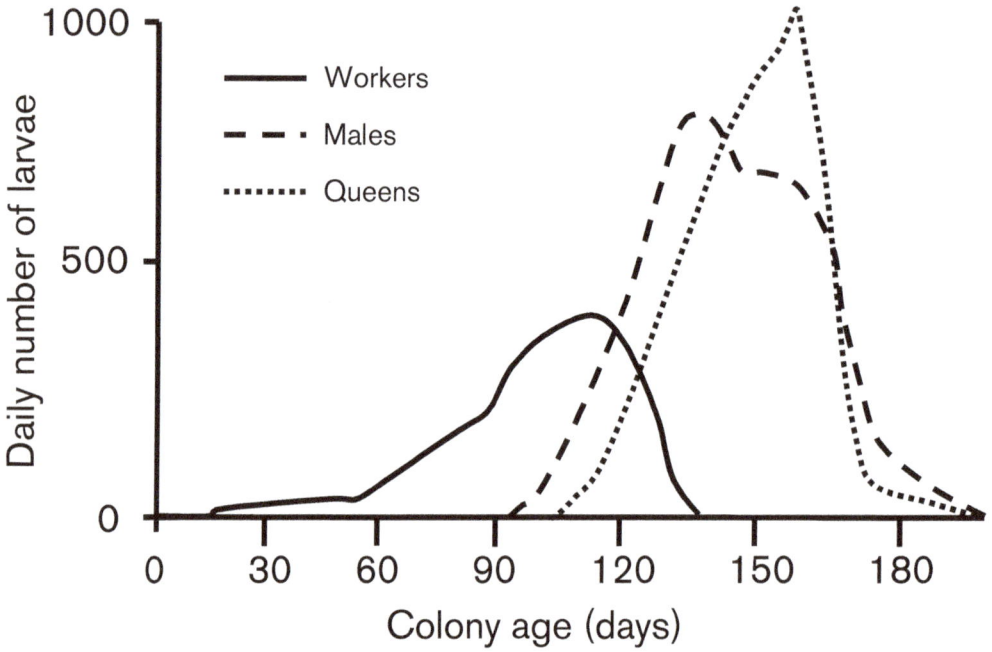

Figure 38. Typical production of workers, then males and finally queens in a long-cycle hornet species like the Asian hornet.

Figure 39. A large mature hornet colony consisting of multiple combs containing a large amount sealed brood (left). It is during this period that the number of workers reach their peak (right). Photographs by S. Martin.

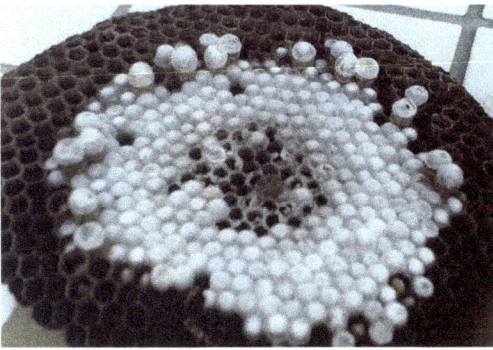

Figure 40. Different brood sized cell silk caps indicate the queen (larger) and male (lower caps with dense white centre). Photograph by S. Martin.

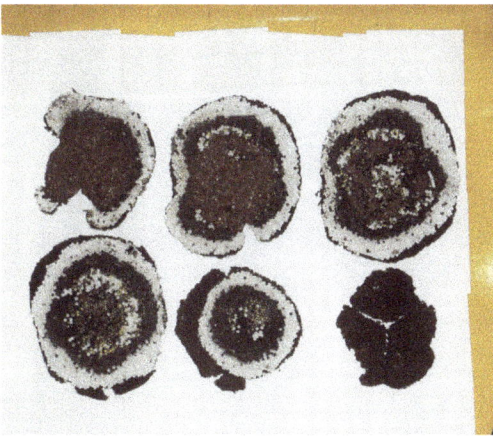

Figure 41. A mature Asian hornet nest consisting of six combs, the sealed brood are contained on the edges of the combs. The centre regions of the top three (oldest) combs have had their empty cells papered over, which is often seen in hornets that build large colonies. Photograph by J. Takahashi.

Mature nests of the European hornet can consist of up to 4,500 cells, whereas, an Asian hornet colony can consist of up to 12,000 cells, contained in up to eleven combs. If you were to open the colony you will see a healthy-looking brood pattern with cells full of eggs, larva at different stages of development and pupae (sealed brood) (Figure 39). The average and maximum number of workers is 182 & 400 in the European hornet, increasing up to 500 & 1400 in the yellow-hornet and 1129 & 1456 for the Asian hornet!

Both the male and queen-destined eggs are laid in some of largest cells built, usually found on the lower combs. In some species of hornets, the size of the silk cap and density of the silk cocoon can be used to determine if the pupae are male or female (Figure 40).

The size range of hornets in a mature nest can be very large since small workers will have been emerged from the upper combs that contain smaller brood

Figure 42. A newly emerged Asian hornet sexual obvious from the pale hairs typical of all recently born hornets, obtaining food from a mature larva. Photograph by J. Takahashi.

cells, while the larger queens and workers will have emerged from the lower combs (Figure 41). Interestingly, male size is largely independent of the size of the cell it was reared in.

After the sexuals emerge from the brood cells they remain in the nest for 8-11 days (males) or 13-14 days (queens). During this time, their weight increases by 40% due to the laying down of fat in the abdomen (Figure 20). These fat stores are built up by eliciting sugars from the larva (Figure 42). This is important since the final few workers are also produced in large cells, so it can be almost impossible to determine based on size alone, if the female is a large worker, or new queen also called a 'gyne'. The only reliable way to determine the caste, queen or worker, is to dissect their abdomen, since new queens build up large fat stores, whereas the workers have none. Once ready to depart, the sexuals leave the colony never to return, so no orientation flight is performed.

Figure 43. Mature sexual hornet larvae that due to constantly supplying their sister queens with nutrition fail to pupate and eventually die within the colony at the end of the season.

Mating

Little is known about the mating behaviour because most species of hornets' mate away from the nest, but scientists think males wait for the queens in areas along certain hedgerows or atop trees. However, the Japanese giant hornet *Vespa mandarina* is an exception with the males waiting outside the nest and chasing the departing queens, capturing her in the air and mating on the ground. Males have also been seen entering the nest and attempting to mate with the queens. Yellow hornet queens seek a hibernation site that can be up to 1.2km away from her colony, although the majority of sites are within a few hundred meters of their original nest.

Colony decline phase (October-November)

This phase is often neglected but is still important, since this is the time when most colonies are found and destroyed. As previously mentioned, as the sexuals emerge they are building up their fat stores, by obtaining nutrients from the larva. However, these sexual larvae are also producing food in the form of a sugary liquid (see diet) to feed the large worker population, as well as attempting to continue their own development from larvae into pupae. So, as autumn progresses and food availability starts to decrease, and the number of newly emerging sexuals continues to increase, a point is reached where the larvae cannot obtain sufficient food for their own development into pupae. Therefore, during the final stages of the colony, when all the pupae have emerged, the colony can contain hundreds of dying mature larva. Many of these are removed by the workers (Figure 43) and piles of dead larvae under or nearby the colony can often be seen. As all the dead larvae were destined to become sexuals under more favourable conditions, this explains why in a prolonged season much greater numbers of sexuals are produced. The workers will continue to use the nest during the night while foraging for other sources of sugars during the day. It is normally the first prolonged frost that finally kills the colony but in areas with warmer climates workers can persist until January before dying of old age. Occasionally a small number of hornets pupate upside down and so deposit their meconium in the top of the cell. When it hardens the adult hornets are unable to escape and become entombed in the cell. These are always useful in species identification of old or abandoned nests. The entire colony cycle for the Asian hornet takes between 8-10 months and just 5-6 months for the European hornet.

Thermoregulation (or heating the colony)

Although insects are normally defined as being cold-blooded, many species have developed amazing ways to actively produce heat for a variety of reasons. The best-known example is the honey bee, but in fact most social insect colonies are able to thermoregulate to some degree. In fact, for the vast majority of flying insects, once they are warmed up to operational temperature, that is warm enough so they can start to fly, losing enough heat generated by their flight muscles to prevent over heating is a major problem. If not, they would soon start to cook themselves from the inside out, since their flight muscles generate large quantities of heat during flying. One of the ways hornets lose heat while flying is they produce a droplet of liquid and hold it between their mandibles during flight. As this evaporates, it removes heat from the hornet, so it is a type of hornet sweating. Most insects when they stop flying cool down to the surrounding air temperature. However, a few species are able to actively generate heat via their powerful flight muscles without flying, as found in the hornets, wasps and honey bees. This allows them to affect the local temperature of their immediate environment, namely their nest. This has many advantages since biological processes, such as development, operate faster at higher temperatures, as can be seen in the drop in the developmental time of the brood as the colony increases in size. Furthermore, having a warm nest allows foraging to occur as soon as there is sufficient light, a period when most of their prey is in a torpid state trying to heat up after the night. This makes foraging very easy and efficient around dawn and dusk. Most importantly, large nests can mount a full defence at any time of the day or night irrespective of the air temperature, as I have so painfully found out. However, thermoregulation is very expensive in terms of energy usage. It is only to be found in social insects since their colonies can consist of tens, hundreds or even thousands of individuals (adults and brood) all living in close proximity. This helps to retain any heat generated by the adults either passively or actively. As mentioned previously (embryo colony) the founding queen of hornets is able to raise the temperature of her embryo nest by curling round the top of the comb, although with no external food stores the amount of heat generated is limited. However the bumblebee queen, who also incubates her first batch of brood, can provide a much longer period of thermoregulation since the queen obtains energy from a nearby cell containing honey. As the number of workers in the colony increase so does the efficiency, and the larvae are able to provide a constant supply of carbohydrates to the thermoregulating workers. By the time the colony has matured and is starting to produce sexuals, the internal nest temperature is maintained day and night at a constant 30 °C (Figure 44). This is possible since the nest wall structure consists of multiple layers of air filled pockets which provide excellent insulation from external temperature fluctuations. For example, the top of large aerial nests is often cone shaped consisting of a very dense structure of air pockets, that protect the colony from both heat and cold as well as rain. The brood and fluid filled sacs act as a massive heat sink and reservoir of heat so increasing the efficiency of any heat the adults generate.

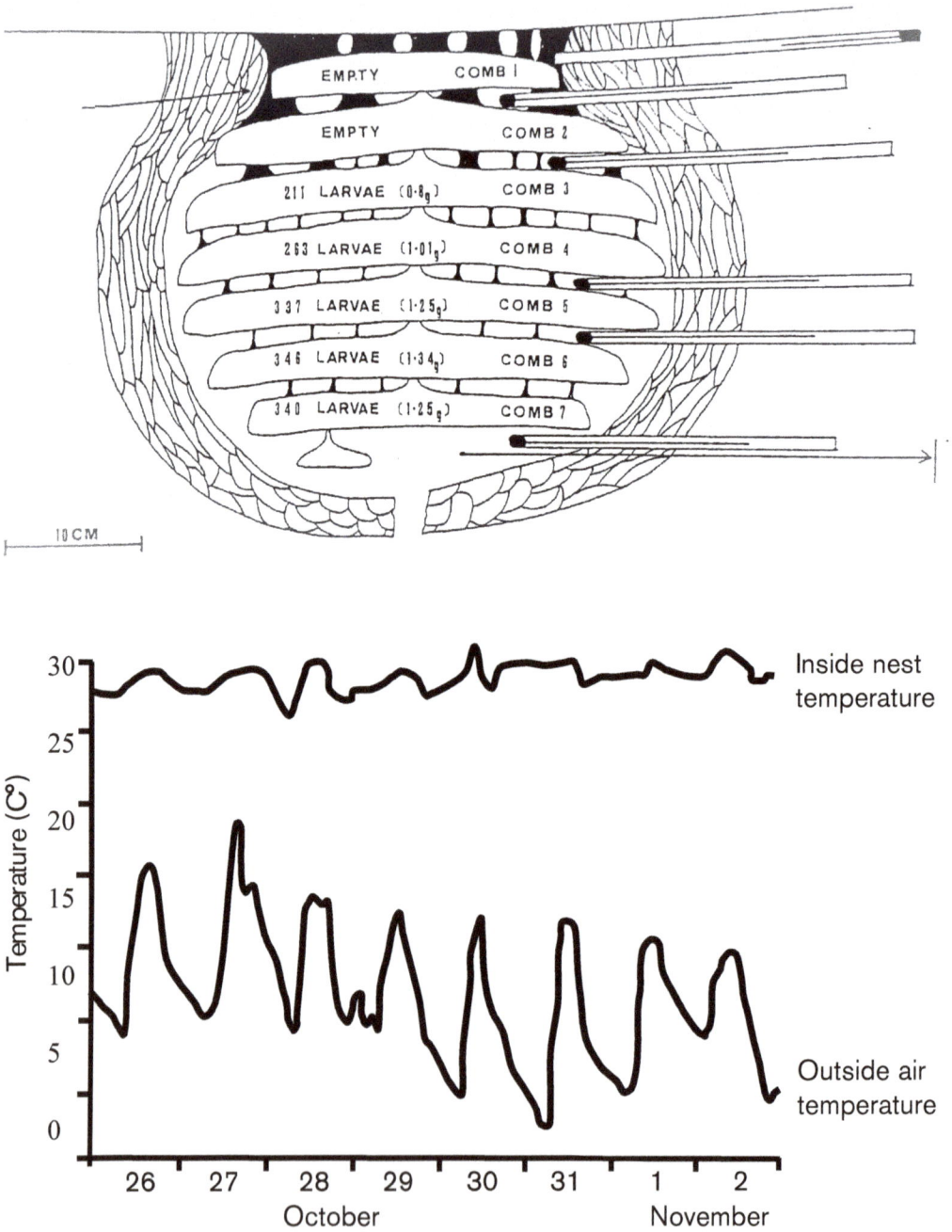

Figure 44. A cross-section through a mature Yellow Hornet colony (upper) showing the heavily insulated walls that allow a constant nest temperature to be maintained (lower) despite large daily changes in the ambient temperature. The diagram shows the position of several thermometers placed into the colony, plus an electronic probe.

*Figure 45. The hornet (*Vespa tropica*) attacking a paper wasp nest. The resident paper wasps are unable to defend the colony and have to leave their colony while the hornet plunders the brood. Photograph by Y. Tatashi.*

Figure 46. Image of Asian hornet (left) and the European hornet (right) feeding on a honey bee. Photographs by C. Luck.

Diet

While bees are strictly vegetarian obtaining a balanced diet from pollen (proteins) and nectar (carbohydrates), wasps are primarily carnivorous (protein), but with a sweet tooth (carbohydrates). The narrow constriction between the thorax and abdomen of all hornets and wasps (petiole) prevents the passage of any solid food (meat) from the mouth into the gut for digestion. This problem is circumvented by refuelling on honeydew, nectar and tree resin while foraging, especially in spring and late autumn, since these are periods when there are no larvae present in the nest. These are valuable sources of carbohydrates (energy) but for the larvae to develop they also require a constant supply of protein. This comes chiefly from preying on other insects. Most hornets are generalists and feed on the whatever is around, plentiful and easily accessible, especially caterpillars, or even stealing food from spider's web. Some hornets have become more specialised, for example, *Vespa tropica* prey exclusively on the larvae of paper wasps (Figure 45). While the little-known species hornet *Vespa binghami* has enlarged ocelli similar to that found in the nocturnal hunting *Provespa* wasps (Figure 4).

The Japanese giant hornet *Vespa mandarinia* will attack weak colonies of other hornet species and some hornets including the Asian hornet have evolved special behaviours to prey on honey bees (see later). When the prey is caught, any indigestible parts like the legs, and head are removed and a meat ball is created from the remaining tissue, often while hanging from a hind leg (Figure 46). During this process, the adults imbibe some of the nutrients from the flesh as they chew the prey before it is fed to the larvae.

On returning to their nest they feed the meat balls to their developing larvae. The larvae are basically a bag with teeth so they don't have any problems digesting solid food, unlike the adults. In return, the larva produces a sugary material that the adult wasps feed on and which sustains all their external activities (Figure 47).

Dr. Takashi Abe of Riken University in Japan analysed the saliva produced by the hornet larvae and discovered it was composed of a selection of 17 to 20 different essential amino acids, and was especially rich in proline, glycine, and alanine. Amino acids are the raw material used to construct proteins, and help hornets to fly 80km a day, non-stop. Dr. Abe took his research further and

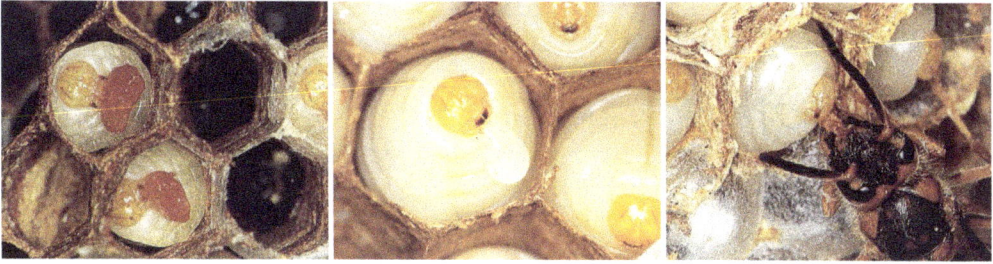

Figure 47. The process of hornet larva feeding on the meat balls (left) provided by the adults. After ingestion, the larvae produce a droplet of sugar rich fluid (centre) that is taken up by the workers to provide energy (right). Photographs by Y. Tatashi.

created VAAM (*Vespa* Amino Acid Mix), which was a synthetic blend of all the amino acids contained in hornet larval saliva. He went on to demonstrate that VAAM helped mice and humans to convert fat into energy more efficiently, which is very important in improving an athlete's performance. This combination of amino acids also helped reduced human fatigue and slowed down the build-up of lactic acid. On these findings VAAM was then commercialised by the Japanese drinks giant 'Meji' and became internationally recognized when Naoko Takahashi won the Sydney Olympic marathon and credited her success to VAAM. Later on, 'running on VAAM-power', Takahashi went on to become the first woman to break the 2 hours 20-minute marathon record, and won six marathons in a row. This drink continues to generated billions of yen each year in Japan for Meji and can be bought from health shops or convenient stores (Figure 48).

Figure 48. The wide range of VAAM products now available to the general public all based on the unique composition of hornet larvae saliva! Photograph by S. Martin.

Chapter 4

Hornet predators and pests

Predators of Hornets

Often when a species is moved accidentally into, or invades, a new environment it becomes free from its natural predators and pests. However, hornets and yellow-jackets, due to their sheer size, painful stings and numerical dominance means they have few natural predators to start with and most of these are highly specialised. In Asia, if a colony is small or weak it can be attacked by other hornets, especially the Japanese giant hornet *Vespa mandarinia*, the world's largest hornet, or the honey buzzard (*Pernis apivorus*), a bird of prey that has specially adapted feathers around its legs to prevent it from getting stung. Occasionally if bears which roam the mountain forests of Asian are particularly hungry they may risk attacking a colony, but this can be hazardous for the bear. Therefore, the largest threat to hornets are humans, who either collect them for food, as occurs throughout most rural areas of Asia, such as the hornet hunters in the Japanese mountains (Figure 49); highly trained pest control offices who kill them where the impact of the hornet colony is a major problem to public health; or people who feel threatened by hornets.

Individual adults can be predated by some bird species, e.g. shrikes, or robber flies (Asilidae) but their impact is very minor, even in the areas were these predatory species are common. In parts of Asia there is also a parasitic cuckoo hornet, which is a rare natural enemy of some species of hornet. The queens of this parasitic species (*Vespa dybowskii*), delay their emergence from hibernation around two months so they are able to invade a nest of their

Figure 49. A good haul of hornet colonies collected in the autumn from the central mountainous region of Japan by a highly skilled 'hornet hunter'. Photograph by S. Martin.

host species such as the yellow or European hornet at the co-operative or early polyethic period. The cuckoo queen has several adaptations, such as a stronger sting and larger mandibles for fighting. Once inside the host's hornet nest it kills the resident hornet queen and becomes the colony's new queen. She uses chemical mimicry to fool the host workers to rear her brood as well as their own. As the cuckoo hornets are a reddish brown in colour and the host species are mainly yellow, you get an unusual change in the colour as the worker population changes from totally yellow to entirely brown as the number of the brown parasitic workers increase over time (Figure 50).

Finally, the cuckoos' sexuals are produced, mate and enter their prolonged hibernation. However, the introduction of this parasitic hornet into Europe as a form of biological control of the Asian hornet would cause many more problems than they would solve. They are one of the most aggressive species I have ever worked with, and in the absence of their host species can revert to establishing their own colony in the normal manner.

*Figure 50. The red parasitic cuckoo hornet (*Vespa dybowskii*) obtaining food from a host yellow hornet worker. Photograph by Y. Tatashi.*

Internal parasites of Hornets

Like all other insects, hornets can become infected by a small number of highly specialised internal parasites. Around 20 species are known from the entire Vespinae family, which reflects the lack of research into this field. Known parasites include Stylopids, nematode worms, moths, beetles, ichneumon flies and some true flies. These groups are all rare and in most cases, are recorded only from yellow-jackets. Very few of these parasites actually kill their hosts, and even fewer actually have a major impact on colony productivity.

Trigonalids wasps are relative large (5-15mm) ichneumonid wasps. They have a unique biology that involves laying thousands of eggs on foliage. These eggs will only hatch if consumed by a caterpillar or sawfly larva. However, the ultimate hosts are the hornet and yellow-jacket larvae, which will become parasitized if they consume an infected caterpillar fed to them by a worker. If more than one trigonalid larva infects the hornet larvae, then cannibalism occurs until only one parasitic larva is left. This larva then devours the pupae and pupates in the sealed cell to emerge as an adult. Several species of hornets, including Asian hornets, have been parasitized by the trigonalid wasps *Bareogonalos jezoensis* (Figure 51).

Stylopids are a highly unusual group of parasitic insects, where the males are free-living having a single pair of fan-like hindwings, since the forewings have been lost (Figure 52). The females remain within the host,

Figure 51. The smaller parasitic trigonalid wasps Bareogonalos jezoensis *developing in the brood cells of the Yellow hornet alongside the hornets. Then emerging from the brood cells in a* Vespa dybowskii *nest (lower left). In* Bareogonalos jezoensis *wasps, the males are much smaller than the females (lower right). Photographs by M. Ono (top) and K. Sayama (bottom).*

Figure 52. The winged male (left) and wingless female (right) stylopids parasites.

enclosed in a puparium, having the form of a larvae. At maturity, the head of the female protrudes from between the segments of the abdomen. The males seek out the females and mate, then the females deposit small first stage larva on flowers that are picked up by other foragers. The larva enters the body and feeding on the body fluids develop within the adults' body. The genus of stylopids that attack hornets and yellow-jackets belong to the *Xenos* group. The number of colonies infected are very small but when they are parasitized, up to 10% of the adults can become infected since nest-mates will be hunting in the same area. The hornets are not usually killed by the parasite although behavioural changes may occur.

Mermithid nematodes belonging to the *Pheromermis* species are a group of specialist's parasites of social insects, particularly hornets and yellow-jackets. Little was known about their biology until New Zealand researchers invested large amounts of resources on understanding their life-history with an aim to using the nematodes as a biological control agent for the yellow-jack-ets. The yellow-jackets had invaded New Zealand in 1944 and again in 1978 and became a major pest, reaching high densities in some areas. The team discovered that the nematode had second intermediate hosts, which were

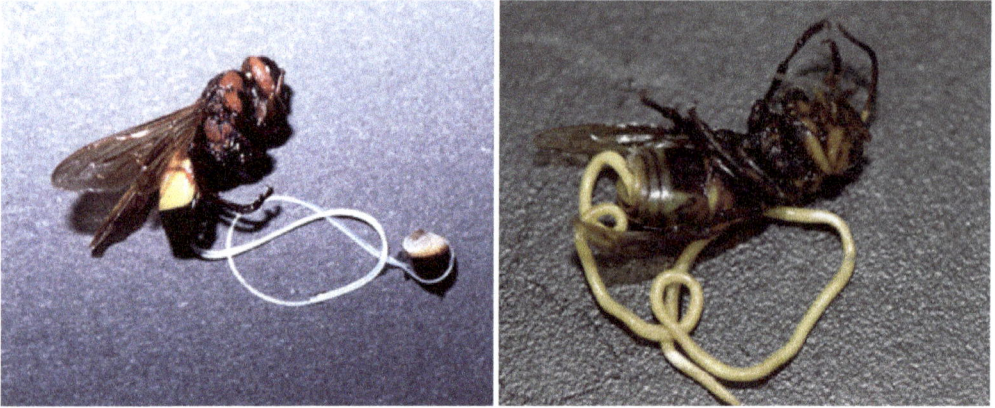

Figure 53. Mature Pheromermis *nematode worms extracted from the abdomen of the hornets* Vespa affinis *(left) and the Japanese Giant hornet (right). Photographs by J. Takahashi.*

freshwater invertebrates. So, if a yellow-jacket captures and feeds on of an infected invertebrate, the nematode develops within the wasp larvae, maturing in the adult. Mature worms entirely fill the yellow-jackets abdomen (Figure 53) and affect their behaviour by forcing the wasp to seek out water. While the wasp drinks the water the worm bursts out of the abdomen and lays eggs near or in the water so completing the life cycle.

In an attempt to control the yellow-jacket populations in New Zealand, millions of artificially reared worms were released into the environment. Despite all these efforts the increase in infestation rates among the yellow-jackets rose from just 3% to 5% and had no impact on the wasp's population that continued to rise. Looking at the biology of this nematode it became clear that it had evolved to only infect workers later in the yellow-jackets colonies life cycle ensuring both parasite and host survived. Subsequently, mathematical models predicted that even if a colony had an 80% level of infection it could still produce sexuals, indicating just how resilient wasp colonies are to infection.

Other highly specialised parasites include the small pyralid moth *Hypsopygia mauritialis* (Figure 54). This is somehow able to enter the hornets' nest, lay its eggs, which develop by feeding on larva and pupae, very much like wax moths in honey bee colonies. There are also conopid flies that wait at the nest entrance and lay their eggs on returning workers, when the larva

Figure 54. A small moth (left) has a similar life cycle to wax moths and the conopid fly (right) a highly specialised parasitic fly whose main hosts are social insects. Photographs by S. Martin (moth) and T. Breistaff (fly).

hatches it develops within the worker killing it when it emerges from its pupae (Figure 54).

Many of these pests and parasites have been investigated as potential biological control agents, but as seen in the case of the nematodes, hornet and yellow-jacket colonies are very resilient against these types of threats. To be really effective the parasite needs to kill the queen or infect the colony at the very early stages of development when the colonies are more vulnerable. However, most of these parasites are naturally rare and normally attack adults in older colonies, so their impact on the wasp population is minimal. This ensures the long-term survival of both the host and parasite.

Chapter 5

Hornets as pests, food & control methods

Hornets and honey bees

The war between honey bees and hornets has been a long evolutionary strug-
gle that continues to amaze and surprise. Any social insect colony such as a
honey bee colony represents a rich and reliable source of food but breaking
into these heavily guarded fortresses is never easy (Figure 55). However, some
species of hornets have evolved some clever adaptions, as have the honey
bees that they predated. A team of researchers showed that even plants
have got involved with the Chinese orchid *Dendrobium sinense* mimicking the
alarm pheromone produced by the Eastern honey bee (*Apis cerana*) in order
to attract the foraging hornet *Vespa bicolor* in order to get pollinated!

*Figure 55. An Asian hornet and
honey bee check each other out,
so see who has the upper hand.
Photograph by C. Luck.*

Hornets attacking Eastern honey bees

The Eastern honey bee (*Apis cerana*) lives in a region where hornets are common and present a constant predation pressure. Therefore, an arms race between hornets and honey bees over millions of years has led to some very interesting behaviour evolving on both sides. Understanding these behaviours helps explain why the Asian hornet is such a threat to Western honey bees. The more agile species of hornets such as the yellow and Asian hornets are able to intercept incoming honey bees in flight when returning to their colony, while other generally larger and less agile species like *Vespa affinis* and the Japanese giant hornet land on the edge of the honey bee colony and dart forward to grab single honey bees. However, regardless of the hornet species, the presence of any hornet in the vicinity of an Eastern honey bee colony elicits the same response. This is when the honey bees move out onto

Figure 56. A Japanese Giant hornet (above) and Asian hornet (lower) engulfed in a ball of Eastern honey bees that eventually results in the death of the hornet. Photographs by S. Martin (Giant hornet) and Y. Sakai (Asian hornet).

the surface around their colony entrance and generate a fizzing-like sound and shimmering behaviour that is due to all the bees simultaneously folding their wings. This type of 'Mexican wave' has a very distinctive sound and visual effect and may be a warning to incoming foragers that hornets are in the area. The other purpose is to maybe warn off the hornet, by way of indicating to the hornet that they have been spotted. However, if a hornet gets too close the Eastern honey bees are able to mount a co-ordinated counter-attack. If a hornet gets too close it can become engulfed in a ball of honey bees within seconds (Figure 56).

Figure 57. A Yellow hornet (top) and Asian hornet (bottom) hawking outside an Eastern honey bee colony, which removes the threat from being balled. Photographs by S. Martin (Yellow hornet) and Y. Sakai (Asian hornet).

As the bees are unable to sting the hornet, due to its thick cuticle (skin), they instead generate heat so the centre of the ball reaches a temperature of around 45°C, which kills the hornet, but only a few bees. However, killing hornets using a thermal ball apparently won't work on the Oriental hornet a species that has a higher tolerance to heat due to being adapted to hot dry environments. However, the Cyprian honey bees still manage to kill the hornet by the ball of bees suffocating the hornet.

To avoid being balled, the more agile hornets like the Asian and Yellow hornets have evolved to 'hawk' incoming bees by positioning themselves several centimetres away from the surface of the colony (Figure 57). Although hornets still will approach bees gathering outside the colony and dart in to try and capture one. Their hawking success rate is low and it can take a long time and many attempts before successfully intercepting an incoming honey bee. It may be that a small number of workers specialise in hawking behaviour. If attacks persist due to the close nature of a hornet colony a colony of Eastern honey bees will often abscond.

Hornets attacking Western honey bees

In Asia, the low level of predation caused by hornets to colonies of Eastern honey bees is not usually regarded as a major problem by beekeepers. However, when Western honey bee (*Apis mellifera*) colonies were introduced into the Far East to increase honey production, hornets soon became serious pests in some areas. The major problem is that the Western honey bee has evolved almost exclusively in regions lacking hornets so that an efficient behavioural response to them has not evolved.

*Attacks by the Japanese Giant Hornet (*Vespa mandarinia*)*

In Japan, this is most dramatically demonstrated when the giant Japanese hornet encountered Western honey bee colonies and quickly learnt that they were unable to defend their colony. When the giant hornet attempts to catch a honey bee it must be very wary of getting caught by the bees. However, Western honey bees retaliate by attacking the hornet but, in an uncoordinated way, with individual bees trying to sting the hornet. This causes no

Figure 58. Mass attack of a European honey bee by the Giant Japanese hornets (top), with the much smaller bees having no chance against the massive hornets (bottom). Photographs by M. Ono.

damage to the much larger giant hornet but causes a dramatic behavioural change in the hornet. So, instead of continuing to prey on single honey bees it starts to recruit other members of its colony. This is done by marking the vegetation around the bee colony with secretions from their sternal glands (Figure 26). The giant hornets then start to arrive at the bee colony and go into an 'attack phase' during which all the bees are killed by the hornets. An entire colony of about 35,000 bees can be killed in just four hours by only 32 giant hornets. The hornets use their massive mandibles to decapitate the bees and all that's left at the end is a big pile of dead bees beneath the hive (Figure 58). After the adult bees have been eliminated, the hornets enter the bee hive and remove the brood and take it back to feed the brood in their colony. It may take several days to remove all the brood from the bee colony, during this period the hornets post guards at the entrance of the bee hive to prevent robbing by other species including other giant hornets from other nearby nests. These mass attacks usually only occur from late august to early November peaking in late September to mid-October, since this is the time when the giant hornet colonies contain the maximum amount of brood, although it remains a mystery why attacks on honey bees do not happen outside this period. Colony losses in some areas can be severe and wire guards have been designed to fit the front of the hives, thus preventing access by the hornets. The foraging distance of the giant hornet is between 1-2 km although up to 8km has been recorded. However, most serious attacks usually are within 1 km of the giant hornet's nest. These mass attacks are dramatic and have been the subject of several natural history films. This behaviour along with their large size make this species popular with the media, who continue to mix up the Asian hornet with the Giant Asian hornet (Figure 58)

Attacks by the Asian hornet (*Vespa velutina*)

The reason why there is so much concern around the spread of the Asian hornet in Europe is primarily down to its ability to prey on honey bees (Figure 59). As previously seen, the Western honey bee has no major adaptions to deal with hornets. Therefore, the hawking success rate of the Asian hornet in Europe has increased dramatically and is compounded by the lack of absconding behav-

Figure 59. The Asian hornet with a captured honey bee (above), however, a large number of bees will be required to feed a large rapidly growing colony (left). Photographs by Y. Sakai (top) and T. Yamamura (bottom).

iour in Western honey bee. This has resulted in up to 40% of the Asian hornet's diet been composed entirely of honey bees in urban areas, a level that falls in forested areas but is still around 20%. Initially, this will only be a local problem by affecting bee colonies within a 1 km of the hornets' nest, but areas up to 3km could be at risk. Although, as colonies start to reach higher densities in an area their impact will increase significantly. This has already been seen in parts of France where hornet nest densities have reached 10 nests per square kilometre. It was estimated by Makoto Matsuura & Seiki Yamane, two famous hornet researchers from Japan that between 30,000 to 60,000 honey bee colonies per year, in Japan are damaged or killed by the predator activities of hornets, which is around 10-20% of all their colonies.

Attacks by the oriental hornet (Vespa orientalis)

In Cyprus and Greece honey bees are under a constant threat of being attacked by the Oriental hornet, *Vespa orientalis* (Figure 60). However, their long association has resulted in a unique defence mechanism. So, like the Asian honey bee (*Apis cerana*) if the hornet ventures close enough a ball of around 100 bees will engulf the hornet (Figure 60) and kill the hornet, although this can take up to an hour. Researchers have shown in this case the bees don't kill the hornet by heating but by asphyxia! Like many large insects the pumping of the abdomen is needed to force air in and out of their breathing tubes. The ball of bees prevents the hornets' abdomen to expand so preventing it from breathing. Furthermore, like most Asian species of honey bees the presence of a hornet near their nest causes the Cyprian honey bee workers to make a hissing noise, although it is unknown if this is a form of communication or just arises due to the stress.

So, it is clear that given enough time natural selection will help the honey bees evolve clever ways of dealing with attacks from hornets. Unfortunately, the hornets are also quick to adapt to new situations and exploit any weaknesses in the honey bees defence system.

*Figure 60. The Oriental hornet (*Vespa orientalis*) attacking a honey bee colony in Greece (left), but get too close and the honey bees are able to ball the hornet and eventually kill it by asphyxiation (right). Photographs by J. Phipps.*

Hornets as stinging insects

Peoples' innate fear of wasps and especially the much larger hornets means a single individual adult can constitute a potential problem. So, hornets have long been feared and for good reason, since their stings are acutely painful and can cause a large amount of swelling (Figure 61). Unlike bees, all hornets and yellow-jackets can sting multiple times as the stingers are not strongly barbed. The only treatment when stung is to wash the area with soap and water and treat with an anti-histamine that *may* reduce the irritation and pain.

Hornets are also very defensive and protect their nest aggressively. The level of aggression varies with many factors, time of year, type of species and perceived level of threat. For example, the Asian hornet in many parts of Asia has a particularly aggressive nature around a mature nest and is greatly feared by the tea pickers up in the highlands of Malaysia. Fortunately, their behaviour of nesting high up in a tree limits the potential interaction between the Asian hornet and humans. Again, in Malaysia a walking group of students climbing the mainland's highest peak 'Mount Tahan' had been chased for almost a kilometre and received numerous stings when they walked unwittingly too close to a colony in a bush, since due to the altitude only low bushes were available as nest sites. There is circumstantial evidence that if you approach a nest too closely you will be given a warning buzz by a worker. If you continue to proceed towards the nest the worker returns to the nest and recruits her sisters to mount a substantial attack.

Throughout Asia numerous people are killed by hornets each year, normally the young and old that accidently disturb a colony. An unlucky few are allergic to the foreign proteins in the hornet venom and as it is with honey bees, one sting is sufficient to set off a potentially lethal allergic reaction. However, due to their large size and ability to deliver multiple stings by tens of hornets, this

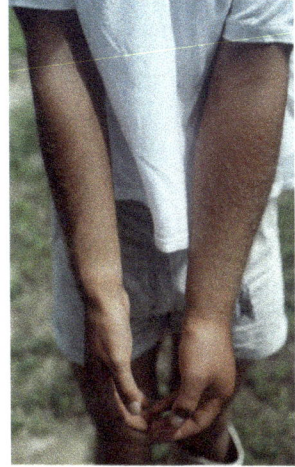

Figure 61. Effect of a hornet sting to the authors arm, the initial pain is severe, but it is the itching later on that is often the worse stage. Photograph by S. Martin

Figure 62. A group of professional hornet researchers removing a mature Asian hornet colony. Three of the researchers are wearing specially designed hornet suits. Photograph by Y. Sakai.

can result in your kidneys shutting down due to toxic overload caused by the sheer amount of venom injected. Therefore, people should always ask for professional assistance if a colony needs to be removed (Figure 62). At least four people have been killed by the Asian hornet in France, most while trying to destroy the nest. In the USA 25% of deaths due to venomous animals were due to yellow-jackets, a similar number due to honey bee stings.

In 2013 hornets killed 42 people and hospitalised over 1000 people in Central China. This was a combination of high numbers of hornet colonies 'a wasp year' and lack of education. As expected the Giant hornet was named as the sole culprit, a ground nesting species, but TV reports showed mature Asian hornet nests high in trees being destroyed. There were many horrifying scenes of people with holes in their arms, which led to the wholesale destruction of all hornet nests including the totally innocent paper wasp and yellow-jacket nests. A key problem, as shown by a Korean study, is that the Asian hornet has become highly adapted to living in urban environments, thus bringing them into direct conflict with humans. The Korean team found that while most hornets occupy either the forest or forest edges the Asian hornet is able to, and actually prefers to nest in urban environments (Figure 63), feeding on the protein waste that we throw away.

This ability to survive in urban environments, nesting in parklands or under the eaves of houses, brings people and hornets into close proximity, which is never a good thing for either party. People in Asian countries are familiar to being around hornets and have learnt not to disturb nests. This lack of knowledge may have contributed to the situation in China where urban expan-

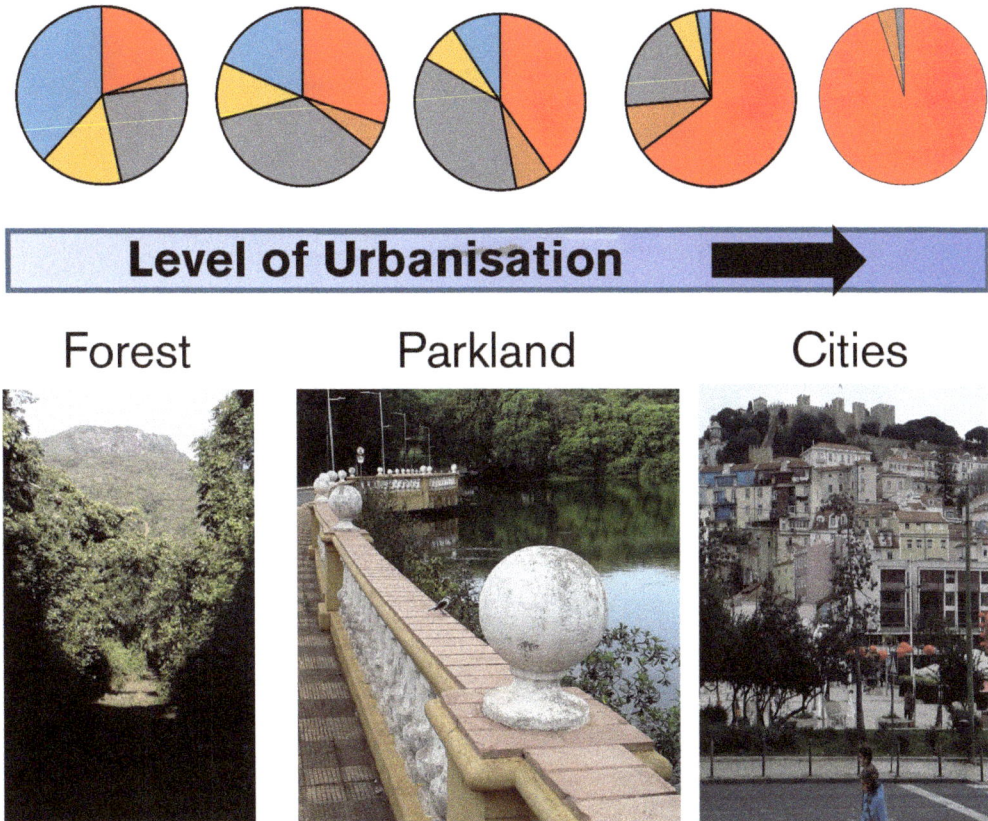

Figure 63. The increasing dominance of the Asian hornet (red) within the hornet communities in Korea in five different habitats that had increasing levels of urbanisation. In forests the Asian hornet, forms part of the hornet diversity along with the yellow hornet (grey), Japanese Giant hornet (yellow), V. analis (orange) & V. tropica (blue). However, as urbanisation increases so does the dominance of the Asian hornet, so cities are almost exclusively home to the Asian hornet. Data adapted from Choi et al., 2012 and photographs by S. Martin.

sion into forested areas is bringing an increased number of people unfamiliar with hornets into their environment. This trend in favouring urban areas has now also been seen in France, where in a region close to the original hornet introduction, densities of Asian hornet nests has reached around 5 nests per square kilometre in rural settings, but this has doubled to 10 nests per square kilometre in urban areas.

Hornets as pests of fruit crops

The impact of hornets in fruit growing areas can be high, since in their search for carbohydrates, especially during autumn, hornets are attracted to fruit just before they are harvested. The hornets are able to remove the outer coat and gnaw at the inside of the fruit and spoil it (Figure 64). This was noticed in the apple orchards in the USA after the European hornet was accidentally introduced. The initial damage also allows other smaller insects access to the inside of the fruit that leads to the fermentation of overripe fruit. In Japan, around 30% of apples were completely eaten by hornets in one area, although the variety had a thin skin, whereas, nearby thick-skinned varieties were largely untouched. Although, in any orchard having hornets flying around while fruit picking is underway is always going to be an additional unwelcome hazard to the harvesters.

Figure 64. Hornets can be a major pest in the fruit industry (left) and damage branches (right) in search for tree resin. Photographs by S. Martin (left) and C. Luck (right).

Hornets as food for humans

Although most people in the west believe hornets are bad, in the east they are viewed in a very different light since they can be an important source of protein in many regions. This is especially true in remote mountainous regions were access to fish may be difficult. In more recent times eating hornet brood and even the adults, is considered a more expensive delicacy. Many high-end restaurants in big cities or specialised rural cafes in Asia offer hornet brood on the menu during the autumn months. In mountain regions of Japan in the farmer cooperative markets combs full of hornet brood can be purchased alongside bottles of hornets soaked in sake (rice wine). In restaurants honey bee larvae or yellow-jacket brood can be mixed into the hornet brood to 'bulk it out'. In small rural restaurants in China the piles of empty hornet combs round the back of the restaurants can be found. There is a small number of highly specialised 'hornet hunters' that collect colonies during the autumn, in order to supply the farmer markets and restaurants. In Japan mature live hornet larvae and pupae are lightly fried in butter with just a dash of salt (Figure 65). They taste like 'warm ice cream'. The adults are deep fried in hot oil, this turns them into tasteless crisps, since the adults consist mainly of chitin. In order to preserve the hornet brood, they are pickled by soaking them in vinegar and soya-sauce, and then eating them any time of the year.

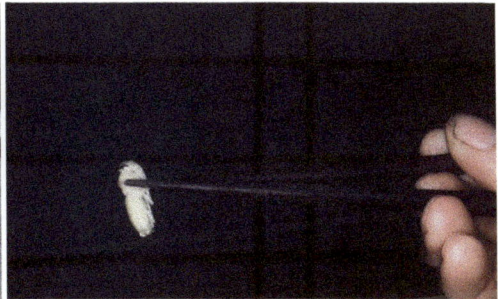

Figure 65. Hornets are still an important source of protein in some areas of Asia, with people enjoying them at dinner parties or in selected restaurants. Photographs by S. Martin.

Hornet control methods

Due to their position at the top of the food chain, hornets have very few natural major predators or pests. The large size of an Asian hornet colony means killing a few individual hornets either physically or using a trap is never going to be effective. In some places, I have seen people either employed or volunteer to protect honey bee apiaries by swatting any hawking hornets (Figure 66). However, unless labour is very cheap this is not effective as the person will need to be employed for months. There are a variety of hornet traps appearing on the market, some are sold as control methods and others under the guise of monitoring. These vary from the basic to more complex (Figure 66). However, evidence from several French studies indicates just how damaging liquid based traps are to the local biodiversity as they all attract a large number of non-target species, including many pollinator species.

The key problem is that workers obtain their sugars from their larva when their colony is active, and its only when the colony is declining after the sexuals are

Figure 66. Hornet control methods range from (Clockwise) swatting hawking workers, the traditional liquid based trap, to more sophisticated designs. Photographs by J. Phipps.

Figure 67. An Asian hornet colony being destroyed using a pesticide foam. Photograph by J. Takahashi.

produced that workers start hunting for sugars and appear in traps. Although it may make you feel good that you are finally capturing the hornets, it's too late to have any lasting effect on the local hornet population. One potential answer is trying to develop a species-specific hornet lure or pheromone trap, which French scientists are trying to develop, but which is notoriously difficult to get right. As we have seen in New Zealand (see section on hibernation), killing queens in the spring would appear to be a logical plan, but again this has almost no impact since queen usurpation is already doing that for you and the natural survival rate of overwintering queens is naturally very low, usually below 1%. Killing overwintering queens has been occurring for 100's of years with Thomas Wildman in 1770 paying a bounty for over-wintering yellow-jacket queens but even back then it didn't work.

The only well-established and proven method of hornet control is colony discovery and destruction (Figure 67), neither of which are easy. Tracking hornets back to their colonies is an old trick used by the hornet-hunters. They capture an individual hornet using a bait trap using raw meat. They then fix a fine cotton thread attached to a small feather around the hornets' waste. The hornet is released and followed by tracking the feather rather than the hornet, which is almost invisible. As hornets returning to their colonies usually go in a straight line this helps locate their colony. Although this needs to be done by professionals since handling hornets is not easy and potentially dangerous. Some people are now looking into using harmonic radar to help track adults in order to locate nests, although this uses highly specialised equipment and is currently very time consuming since each hornet needs to be fitted with a metal antenna! Once the colony is, located normally high up in a tree, its destruction is again not an easy matter and should only be done by professionals (Figure 68).

Figure 68. Crane lifts two professionals on a platform high up into the canopy to remove an Asian hornet colony. Photographs by J. Takahashi.

This can be done using fire, but this is not advisable, better to inject insecticides into the colony. If possible, working at night ensures as many of the workers are killed as possible, but this is often not practical or safe. The most efficient way is to remove the entire colony if possible as it allows a large amount of key biological data about the hornets, such as the timing of sexual production, which is especially important when colonies in new locations are found. The resilience of hornets and the problem of effective control of Asian hornet populations was recently demonstrated by a team of French mathematical modelers headed by Christelle Robinet that predicted if 60% of colonies were controlled (killed) subsequent spread would be reduced by only 17% and nest densities by 28%. Even if a very unrealistic 95% of colonies were destroyed, this would only reduce both their subsequent spread and nest density by only 50%.

People are coming up with ingenious new ways of killing colonies, such as developing drones to inject pesticides into the nests and no doubt many more will appear. However, the most important information is never attempt to remove or kill an Asian hornet colony, since if things go wrong, and they do, then you could be in serious trouble as evidenced by professionals in France being killed during the removal of Asian hornet colonies. Always contact the relevant authority in your country which will then arrange for the appropriate people to be contacted. For example, in the UK if you find a nest or see Asian hornets hawking your bees, take a photograph and use the 'Asian hornet watch app from the 'Centre for Ecology and Hydrology' or send a good quality photograph to the National Bee Unit and they will arrange for qualified people to track and destroy the colony. As the situation changes in each country, as the Asian hornet becomes established, so will the policy and number of people qualified to deal the hornet reports.

Protecting your honey bee colonies

Protecting your honey bees from hornet predation has never been, nor will it be easy. In Japan wire screens placed over the front of hives prevents the Japanese Giant hornet from gaining entry. In areas where hornets are common, local beekeepers often use hives with very narrow entrances (Figure 69). However, this only prevents the entry to the hive, which is not a major issue with the Asian hornet, since the main predation occurs outside the hive.

However, this is of limited use against the Asian hornets since their preferred hunting method is the hawking of returning honey bees, not trying to gain entry to the hive. It is not really effective to kill individual hawking hornets as more will arrive, so getting rid of the colony is the best way to protect your honey bees.

On the market, several hornet traps are being advertised, from a plastic pretend wasp nest claimed to scare other wasps (hornets?) away! Some like

Figure 69. Traditional Eastern honey bee hives on the small island of Tsushima that have a very narrow bee width entrance to prevent hornets entering the colonies. Photograph by J. Takahashi (top) and Y. Sakai (bottom).

ApiShield@, which is a hive floor has at least some logic behind its design and since it does not use a bait, non-target organisms such as pollinators are not affected. Low tech traps made from plastic drink bottles and containing a variety of baits depending on the time of year are being promoted as being effective, although as previously mentioned they normally end up killing more non-target organisms rather than Asian hornets. Often traps provide the illusion that they are having an impact, but the sheer size of the Asian hornet colonies and nesting densities they can reach means the actual impact is always much lower than expected.

For example, a French trial evaluated the various Asian hornet trapping methods available on the French market. These were a combination of traps or domes, combined with home-made sweet attractant, home-made protein attractant, home-made wax based attractant, and a Véto-pharma trap. They were tested in apiaries (August to November) in autumn near Toulouse in France where there is a high infestation of Asian hornet. Even under these optimal conditions that guaranteed very large numbers of Asian hornet in the area, the maximum number of hornets trapped per day was just ten. This will be just a fraction of the number that would be dying daily from natural causes, while between 80% to 100% of the insects in the traps were non-target organisms! As the Asian hornet spreads and more funds are put into developing specific baits things may improve, but it will only be when they become a real concern for the general public that things will change.

The future

Given the environmental conditions of the native range of the Asian hornet (high mountains to lowland tropical wetlands) means that there are few if any natural barriers to prevent the current spread of the hornet. Its slower spread in Korea may be down to competition with other hornets, since their numbers have become greatly reduced as the Asian hornet population has become established, especially in urban areas. The constant supply of widely available food such as honey bees, may provide a stable and long-lasting food resource that allows colonies to produce high number of sexuals in less than favourable weather conditions (Figure 70). The lack of competition for nest sites, reduces queen usurpation, and increases the success rates of the founding queens. Over time the density of colonies will increase and specialised pest control companies will benefit. However, my concern is for all the other species caught up in the attempts to control the Asian hornets. Already hornet traps are being widely sold and advocated despite the scientific evidence that they do more harm than good. Countless native hornet and wasp colonies will be destroyed and millions of insects, especially bees and flies, will be killed. The hornets have enough impact on our fragile biodiversity and we should be careful not to add to the Asian hornet's impact. Over time the western honey bees will adapt to this new threat, but in the meantime, we all need to consider our actions on the wider pollinator community as we also adapt to the presence of the Asian hornet.

Figure 70. A massive two-meter long Asian hornet colony collected from Tsushima Island by Junichi Takashi a Japanese hornet expert. Photographs by J. Takahashi.

Endpiece

Asian Hornet Hawking by Y. Sakai

Asian Hornet Hawking by Y. Sakai

Further reading

Further reading (books)

Archer ME (2012) Vespine Wasps of the World; Behaviour, Ecology & Taxonomy of the Vespinae. Siri Scientific Press.

Edwards R (1980) Social wasps: Their biology & control. The Rentokil library, East Grinstead.

Hunt JH (2007) The Evolution of Social Wasps. Oxford University Press.

Matsuura M, Yamane S. (1984) Biology of the Vespine wasps. Springer-Verlag.

Spradbery JP (1973) Wasps: An account of the biology and natural history of solitary and social wasps. Sidgwick & Jackson, London. ISBN 0 906527 15 5

Further reading (scientific papers)

PREFACE

Bumblebee Economics. By Bernd Heinrich. Harvard University Press (2004) ISBN 9780674016392

Martin SJ (1988) Thermoregulation in *Vespa simillima xanthopthera*. Kontyu, 56, 674-677.

Chapter 1. GENERAL INTRODUCTION TO HORNETS & YELLOW-JACKETS

Archer ME (2012) Vespine Wasps of the World; Behaviour, Ecology & Taxonomy of the Vespinae. Siri Scientific Press.

www.BWARS.com The Bees, Wasps and Ants Recording Scheme, contains lots of information on the British species and distribution maps.

Carpenter JM (1982) The phylogenetic relationships and natural classification of the Vespoidea (Hymenoptera). Systematic Entomology, 7, 11-38.

Carpenter JM, Kojima J (1997) Checklist of the species in the subfamily Vespinae (Insecta: Hymenoptera: Vespidae). Natural history bulletin of Ibaraki University, 1, 51–92.

Chauzat M-P, Martin SJ (2009) A foreigner in France, Biological information on the Asian hornet *Vespa velutina*, a recently introduced species. Biologist, 56, 86-91.

Martin SJ (1995) Hornets of Malaysia. Malayan Nature Journal, 49, 71-82.

Perrard A, Pickett K. M, Villemant C, Kojima J-I, Carpenter J (2013) Phylogeny of hornets: a total evidence approach (Hymenoptera, Vespidae, Vespinae, *Vespa*). Journal of Hymenopteran Research. 32, 1–15. doi:10.3897/JHR.32.4685

Chapter 2. HORNET & YELLOW-JACKET INVASIONS

Arca M, Mougel F, Guillemaud T et al (2015) Reconstructing the invasion and the demographic history of the yellow-legged hornet, *Vespa velutina*, in Europe. Biological Invasions 17, 2357. doi:10.1007/s10530-015-0880-9

Barbet-Massin M, Rome Q, Muller F, Perrard A, Villemant C, Jiguet F (2013) Climate change increases the risk of invasion by the Yellow-legged hornet. Biological Conservation, 157, 4–10

Beggs JR, Brockerhoff EG, Corley JC et al (2011) Ecological effects and management of invasive alien Vespidae. Biocontrol, 56, 505–526.

Choi MB, Martin SJ, Lee JW (2012) Distribution, spread, and impact of the invasive hornet *Vespa velutina* in South Korea. Journal of Asia-Pacific Entomology. 15, 473–477.

Choi MB, Lee S-A, Suk HY, Lee JW (2013) Microsatellite variation in colonizing populations of yellow-legged Asian hornet, *Vespa velutina nigrithorax*. South Korea. Entomological Research, 43, 208–214

Franklin DN, Brown MA, Datta1 S, Cuthbertson AGS, Budge GE, Keeling MJ (2017) Invasion dynamics of Asian hornet, *Vespa velutina* (Hymenoptera: Vespidae): a case study of a commune in south-west France. Applied Entomology and Zoology, 52, 221–229. DOI 10.1007/s13355-016-0470-z

Kim JK, Choi M, Moon TY (2006) Occurrence of *Vespa velutina* Lepeletier from Korea, and a revised key for Korean *Vespa* species (Hymenoptera: Vespidae). Entomological Research, 36,112–115. doi:10.1111/j.1748-5967.2006.00018.x

Milanesio D, Saccani M, Maggiora R, Laurino D, Porporato M (2016) Design of a harmonic radar for the tracking of the Asian yellow-legged hornet. Ecology and Evolution, doi: 10.1002/ece3.2011

Moller H (1996) Lessons for invasion theory from social insects. Biological Conservation, 78, 125–142.

Monceau K, Arca M, Leprêtre L, Mougel F, Bonnard O, Silvain J. F, Maher N, Arnold G, Thiéry D (2013) Native prey and invasive predator patterns of foraging activity: the case of the yellow-legged hornet predation at European honeybee hives. PLoS ONE 8(6):e66492.

Robinet C, Suppo C, Darrouzet E (2017), Rapid spread of the invasive yellow-legged hornet in France: the role of human-mediated dispersal and the effects of control measures. Journal of Applied Ecology, 54, 205–215. doi:10.1111/1365-2664.12724

Takeuchi T, Takahashi R, Kiyoshi T, Nakamura M, Minoshima YN, Takahashi J (2017) The origin and genetic diversity of the yellow-legged hornet *Vespa velutina* introduced in Japan. Insectes Sociaux, doi:10.1007/s00040-017-0545-z

Mikheyev AS, Mandy MMY, Arora J, Seeley TD (2015) Museum samples reveal rapid evolution by wild honey bees exposed to a novel parasite. Nature Communications, 6, 7991.

Rome Q, Dambrine L, Onate C, Muller F, Villemant C, Garcia-Perez L, Maia M, Carvalho Esteves P, Bruneau E (2013) Spread of the invasive hornet *Ves-*

pa velutina Lepeletier, 1836, in Europe in 2012 (Hym., Vespidae). Bulletin de la Société entomologique de France, 118, 21–22.

Villemant C, Haxaire J, Streito JC (2006) Premier bilan de l'invasion de *Vespa velutina* Lepeletier en France (Hymenoptera, Vespidae). Bulletin de la Société entomologique de France, 111, 535-538.

Chapter 3. HORNET LIFE CYCLE

Mikkola K (1984) Migration of wasp and bumble bee queens across the Gulf of Finland (Hymenoptera: Vespidae and Apidae). Notulae Entomologicae 64, 125-128.

Martin SJ (1990) Nest thermoregulation in *Vespa simillima*, *V. tropica* and *V. analis*. Ecological Entomology, 15, 301-310.

Martin SJ (1991) A simulation model for colony development of the hornet *Vespa simillima* (Hymenoptera, Vespidae). Japanese Journal of Entomology, 59, 105-124.

Martin SJ (1992) Colony defence against ants in *Vespa*. Insectes Sociaux, 39, 99-113.

Martin SJ (1993) Weight changes in adult hornets, *Vespa affinis*. Insectes Sociaux. 40, 363-368.

Matsuura M, Yamane S. (1984) Biology of the Vespine wasps. Springer-Verlag.

Chapter 4. HORNET PREDATORS & PESTS

Brodmann, J, Twele R, Francke W, Yi-bo L, Xi-qiang S. Ayasse M. (2009). Orchid mimics honey bee alarm pheromone in order to attract insects for pollination. Current Biology, 19, 1368-1372.

Martin SJ (1988) Occurrence of *Anthrax distigma* (Diptera, Bombyliidae) in the nest of *Vespa simillima xanthoptera* (Hymenoptera, Vespidae). Kontyu, 56, 461-462.

Martin SJ (1992) Occurrence of the Pyralid Moth *Hypsopygia mauritialis* (Lepidoptera, Pyralidae) in the nests of *Vespa affinis*. Japanese Journal of Entomology, 60, 267-270.

Sayama K, Kosaka H, Makino S (2013) Release of juvenile nematodes at hibernation sites by overwintered queens of the hornet *Vespa simillima*. Insectes Sociaux, 60, 383-388.

Chapter 5. HORNETS AS PESTS, FOOD & CONTROL METHODS

Beggs JR, Brockerhoff EG, Corley JC, Kenis M, Masciocchi M, Muller F, Rome Q, Villemant C (2011) Ecological effects and management of invasive alien Vespidae. BioControl, 56, 505–526.

Ken T, Hepburn HR, Radloff SE et al (2005) Heat-balling wasps by honeybees. Naturwissenschaften, 92, 492–495.

Kishi S, Goka K (2017) Review of the invasive yellow-legged hornet, *Vespa velutina nigrithorax* (Hymenoptera: Vespidae), in Japan and its possible chemical control. Applied Entomology & Zoology, 1-8. doi:10.1007/s13355-017-0506-z

Landolt P, Zhang QH (2016) Discovery and development of chemical attractants used to trap pestiferous social wasps (Hymenoptera: Vespidae). Journal of Chemical Ecology, 42, 655–665. doi:10.1007/s10886-016-0721-z

Martin SJ (2004) Biological control of social wasps (Vespinae) using mermitid nematodes. New Zealand Journal of Zoology, 31, 241-248.

Ono M, Igarashi T, Ohno E, Sasaki M (1995) Unusual thermal defence by a honeybee against mass attack by hornets. Nature, 377, 334–336. doi:10.1038/377334a0

Papachristoforou A, Rortais A, Zafeiridou G, Theophilidis G, Garnery L, Thrasyvoulou A, Arnold G (2007) Smothered to death: Hornets asphyxiated by honeybees. Current Biology, 17, R795-796.

Sayama K, Kosaka H, Makino S (2007) The first record of infection and sterilization by the nematode *Sphaerularia* in hornets (Hymenoptera, Vespidae, *Vespa*) Insectes Sociaux, 54, 53-55. doi:10.1007/s00040-007-0912-2

Tan K, Radloff S, Li JJ, Hepburn HR, Yang MX, Zhang LJ, Neumann P (2007) Bee-hawking by the wasp, *Vespa velutina*, on the honeybees *Apis cerana* and *A. mellifera*. Naturwissenschaften, 94, 469–472.

Villemant C, Barbet-Massin M, Perrard A et al (2011) Predicting the invasion risk by the alien bee-hawking yellow-legged hornet *Vespa velutina* nigrithorax across Europe and other continents with niche models. Biological Conservation, 144, 2142–2150.

An Asian hornet hawking in front of an Apis cerana colony in a traditional hive on Tsushima Island, Japan. Photograph by Y. Sakai.

Asian hornet resting on flower in France. Photograph by L. Naiff.